AQA GCSE SCIENCE FOUNDATION

REVISION AND PRACTICE BOOK

Editor: Steve Witney

Ian Brandon, Barbara Drozdowska, Joyce Gustard,
Terry Mansfield, Christine Woodward

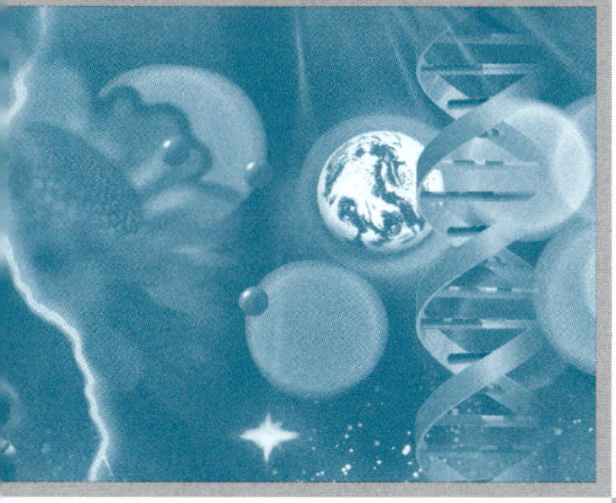

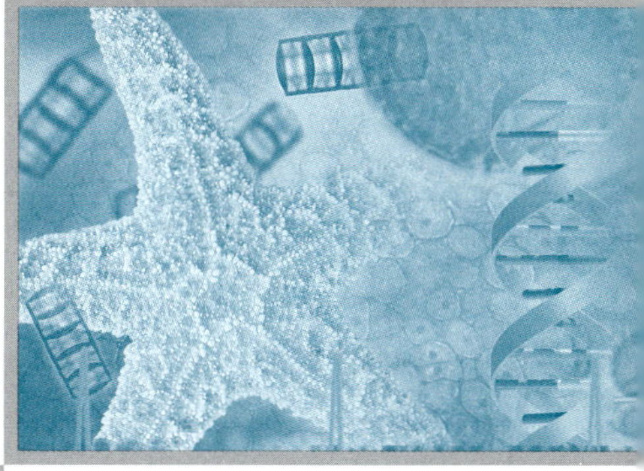

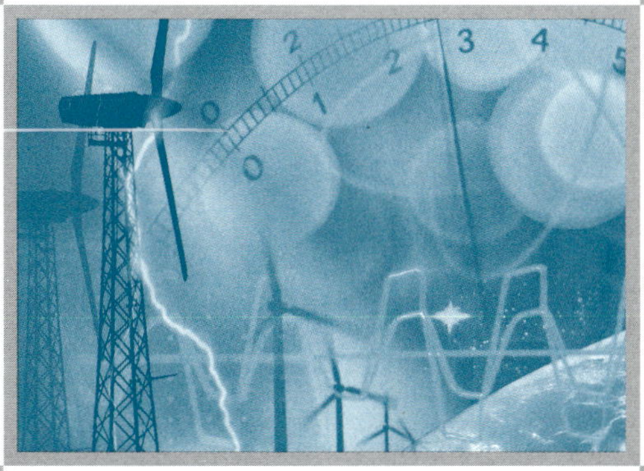

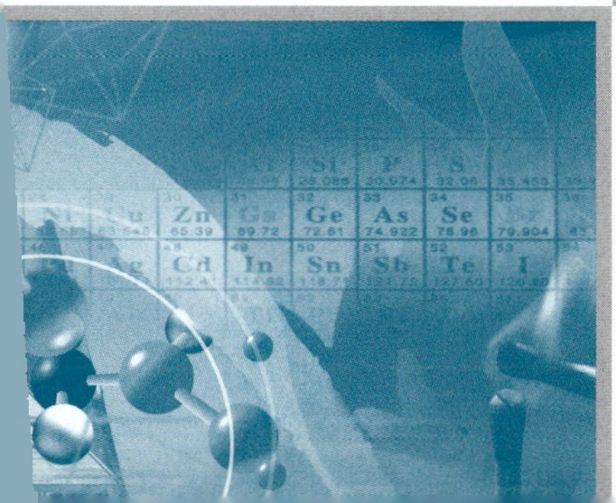

Hodder & Stoughton
A MEMBER OF THE HODDER HEADLINE GROUP

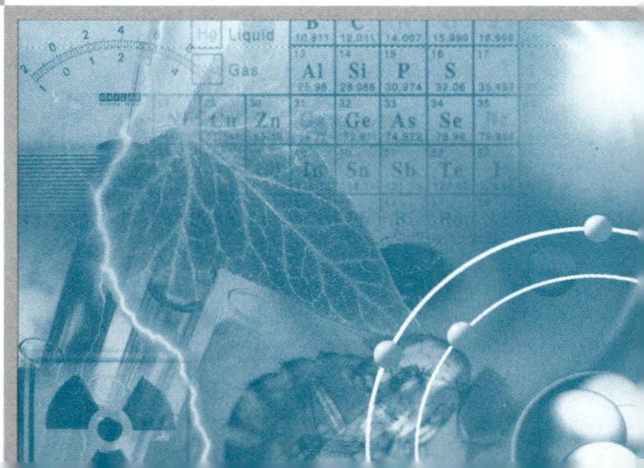

Acknowledgements
The publishers would like to thank the following company for permission to reproduce photographs in this book.

Corbis (page 35 top left, both bottom), Silkeborg Museum, Denmark (page 35 top right)

The publishers would also like to thank AQA for kind permission to re-use their examination questions.

Orders: please contact Bookpoint Ltd, 130 Milton Park, Abingdon, Oxon OX14 4SB. Telephone: (44) 01235 827720. Fax: (44) 01235 400454. Lines are open from 9.00–6.00, Monday to Saturday, with a 24 hour message answering service. Email address: orders@bookpoint.co.uk

British Library Cataloguing in Publication Data
A catalogue record for this title is available from the British Library

ISBN 0 340 81303 2

First published 2004
Impression number 10 9 8 7 6 5 4 3 2 1
Year 2009 2008 2007 2006 2005 2004

Typeset by Tech-Set Ltd. Printed in Spain for Hodder & Stoughton Educational, a division of Hodder Headline, 338 Euston Road, London NW1 3BH.

Contents

Specification matching summary

The contents of this book are designed to cover all aspects of the knowledge and understanding needed if you are taking either the AQA Modular or the AQA Co-ordinated GCSE Double Award Science examination.

The contents are divided into chapters that match the organisation of the Modular course. However, each block of four modules covers one section of the Co-ordinated course and the core content of one of the separate science GCSE courses in: Biology, Chemistry, Physics, and Biology (Human).

Modules	Co-ordinated Science	Separate Science
1, 2, 3, 4	Life Processes and Living Things	Biology/Biology (Human)
5, 6*, 7, 8	Materials and their Properties	Chemistry
9, 10, 11, 12	Physical Processes	Physics

*Part of Module 6 – 'The Structure of the Earth' – is also examined within the Co-ordinated Science *Physical Processes* examination paper and within the separate Physics examination (both Co-ordinated and Modular).

How to use this book

This book can be used throughout your science course in addition to forming the basis of your science revision programme before your examinations. It contains all the information you need to know in brief easy to follow notes and diagrams. Key terms are highlighted in blue.

EXAMINER'S TIP

The book includes a number of useful Examiner's tips to help you understand and remember the work you have covered for your GCSE examination. Read these carefully to see how the advice relates to the topic you are revising.

Throughout each module are short Check Your Understanding questions. These are not actual examination questions, but give you a quick check on the progress you are making. Answers are given at the back of the book.

At the end of each module is a complete Topic Test. Before completing a topic test you may need to go back and read through the module again, this will help you to remember the facts and ideas. The answers provided at the back of the book have been broken down to show you how marks may be awarded in the final examination. Completing the tests will help you to highlight both the strong and weak areas in your knowledge and understanding. If you score very low marks for a topic test you will at least have identified an area on which you need to spend more time.

At the end of the biology, chemistry and physics sections are two examination questions taken from recent AQA examinations, together with model answers. The answers have been annotated with useful examiner's comments, hints and explanations. Read through the questions and see if you would have given the same answers. Check how and why the marks have been awarded. Read and take note of the hints and apply these to when you do practice questions and of course the real thing!

A number of key terms, together with their meanings, are given in the combined Glossary/Index on pages 156 to 167. Use the glossary as a checklist – do you know the meaning of each of the terms? If not, then learn them. As a test, ask a friend to pick a few words for you to give the meanings.

Equations and units.

There are a number of equations, particularly in the physics modules, that must be learnt by heart. These have been written inside a box with a blue border. Learn the words, not just the symbols. Remember also to learn the UNITS that go with each QUANTITY.

If you are following the Modular science course, modules 1, 2, 5, 6, 9 and 10 will be examined by End of Module Tests. The tests will be taken at appropriate times during the course, probably in Year 10. But don't ignore these modules when you are revising for your final examinations at the end of Year 11. Parts of the modules, highlighted in the text by a blue strip down the left hand side, may also be examined in the terminal papers.

Help with your science revision

Organising yourself

Give yourself plenty of time to revise the subject thoroughly. Have a plan, don't leave it until the last moment. Remember that if you are taking the modular course you will probably have some of your GCSE exams in Year 10. These need to be revised for too.

Work out how many days you have until your exams and how many hours you can realistically spend on revision. Use this information to plan an efficient revision timetable. Remember, add in some leisure activities, you can't be revising all of the time.

Break your revision up into slots of short 30 minute sessions with a break after each session. For most people little and often is better than trying to do everything in one go. After your break start the next revision session by quickly reviewing what you revised in the previous session. This helps to fix it in your mind. Make sure you give yourself a little reward for completing a session and achieving a target.

Find somewhere quiet and comfortable to work, if possible with a desk or table. Make sure you are away from distractions like the TV, or younger brothers and sisters. Have pens, pencils, highlighter pens, notepads, your textbook, notes, past papers and this revision book ready to use. You might not need them all at once, but it saves time if they are all readily available.

As you revise, the list of topics you feel you understand well will grow and your list of weaker topics will get smaller (hopefully!). Leave some extra time closer to the exams to go through and sort out your weaker areas. If you are really struggling with a concept or topic ask your science teacher for help.

Pulling it all together

Don't just revise a topic only once – it won't stick. Go back to the topic as frequently as you can. Don't just read through this book, do something active. Highlight the key points with colours and symbols. Rewrite sections from the book as a series of bullet points. Try the questions, and ask friends or family to test you. If possible record key points, equations and definitions on a tape, then play the tape back frequently.

'Go for 3'

This is a simple technique where you write down key words and then make a list of at least three important things about them. Examination questions rarely expect you to give more than three pieces of information. For example:

atom

1 The smallest part of an element.
2 Nucleus contains protons and neutrons.
3 Electrons orbit the nucleus.

proton

1 Found in the nucleus of an atom.
2 Has a single positive charge.
3 Has a relative mass of 1.

A letter sequence/mnemonic

Make up a letter sequence, or series of words, to help remember an important fact, for example AAA = All Arteries carry blood Away from the heart.

A mind map/spider diagram

You can use a mind map to summarise a topic and show the links between the ideas that you need to learn. You could use a spider diagram to ask yourself those important questions that you know you need to be able to answer. For example:

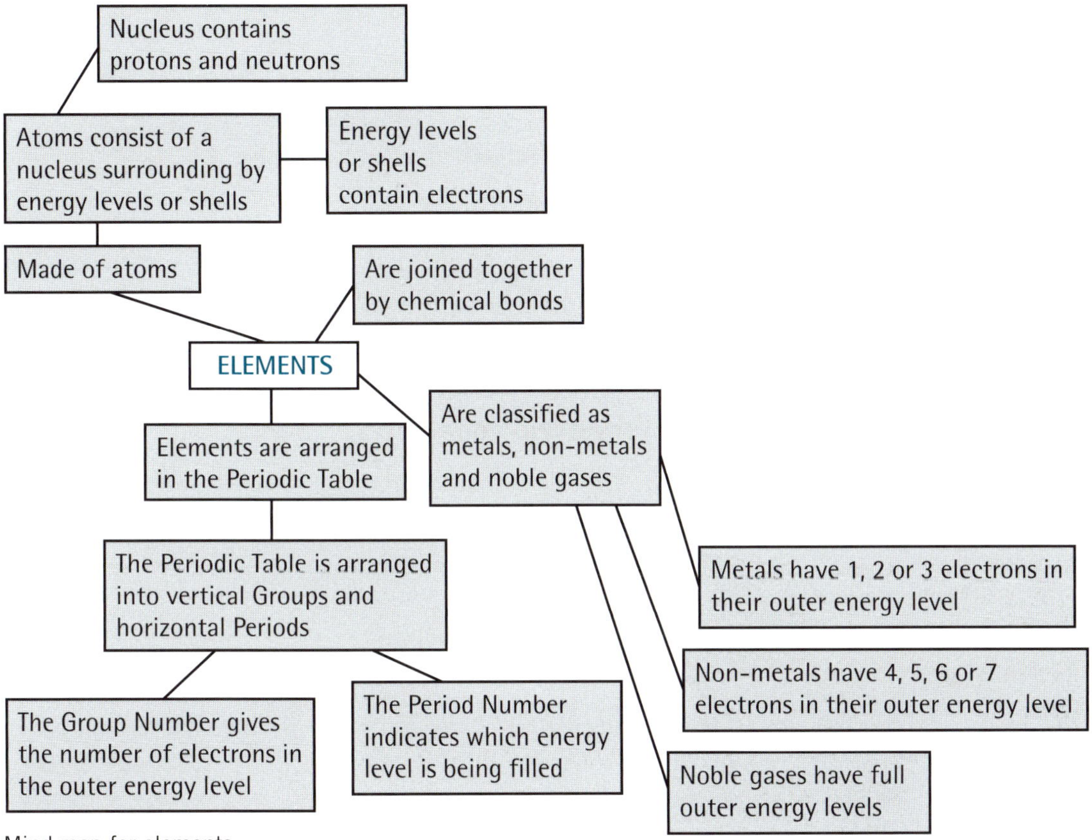

Mind map for elements

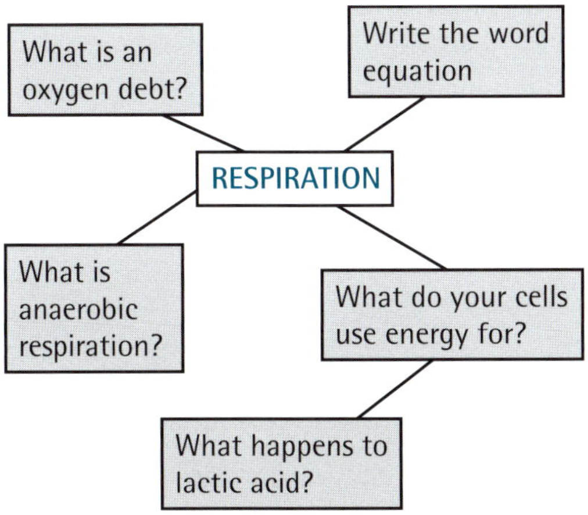

Spider diagram for respiration

Key fact cards

Choose part of a topic. On one side of a postcard write down four or five key facts about that topic. Keep the cards in a pile. For example:

> **Parallel circuits**
>
> Potential difference across each component is the same.
>
> The bigger the resistance of a component the smaller the current flowing through it.
>
> The total current flowing = sum of the currents through each component
>
> $$I_{total} = I_1 + I_2 + ...$$

A key fact card for parallel circuits

When you have a few minutes to spare learn one or two of the cards. Ask someone to pull a card from the pile and test you. If you can answer their questions put the card at the bottom of the pile. If you can't put the card at the top and learn it next time you have a few spare minutes.

Annotated diagrams

Draw simple diagrams, pictures, graphs or tables and add your notes, labels or instructions. For example:

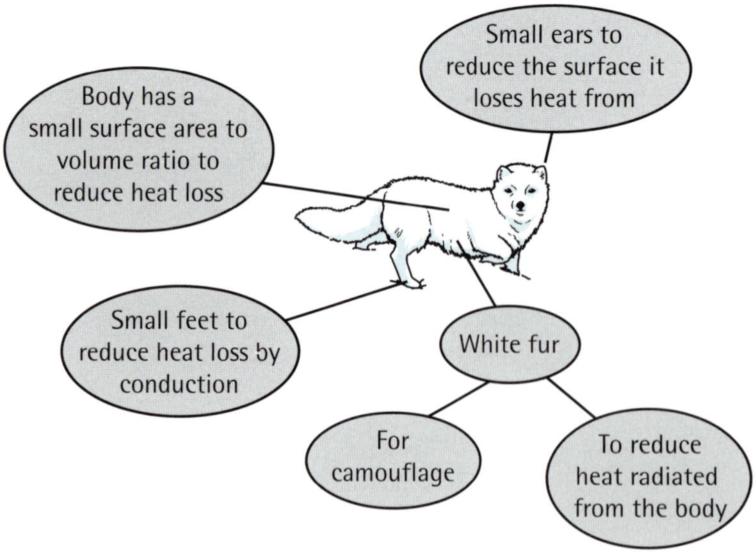

An annotated diagram about arctic foxes

HUMANS AS ORGANISMS

Cells

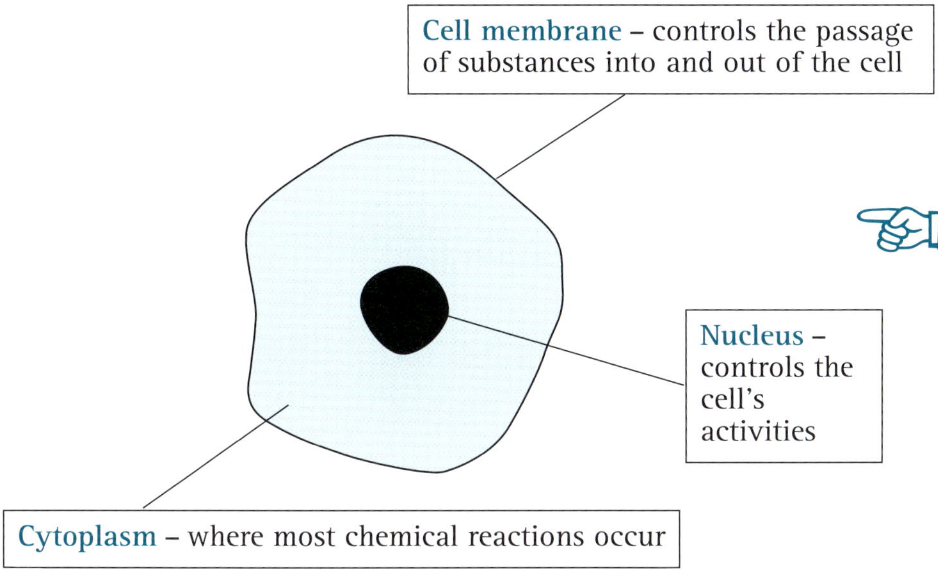

Figure 1.1 A typical animal cell

You must be able to LABEL THESE PARTS on a diagram and STATE THEIR FUNCTIONS.

Some cells look different from this one but still contain the same parts. These cells have adapted to fulfil a particular role in the body.

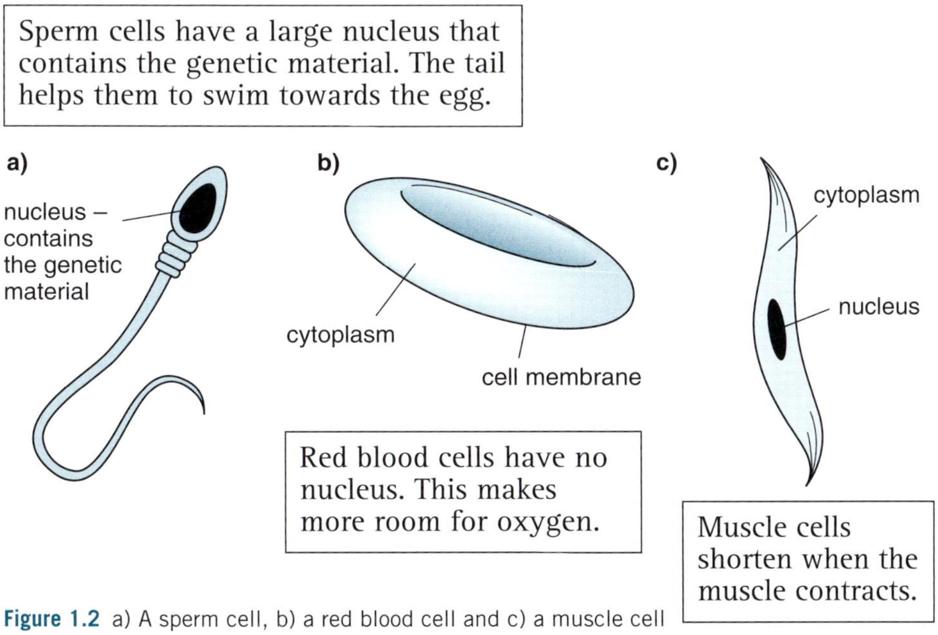

Figure 1.2 a) A sperm cell, b) a red blood cell and c) a muscle cell

How do dissolved substances get into and out of cells?

Diffusion

Diffusion is the movement of molecules of a gas or any substance in solution from where they are in HIGHER concentration to a region where they are in LOWER concentration.

LEARN this definition.

Diffusion will be faster if:

- the differences in concentration are large
- the surface area of the membrane is large.

In the alveoli in the lungs the concentration of oxygen is always large because we are breathing in more oxygen all the time. So oxygen always diffuses into the blood (see Figure 1.7 on page 5). The surface area of the lungs is increased by the alveoli.

Carbon dioxide diffuses from blood into air.

Oxygen diffuses from air into lungs.

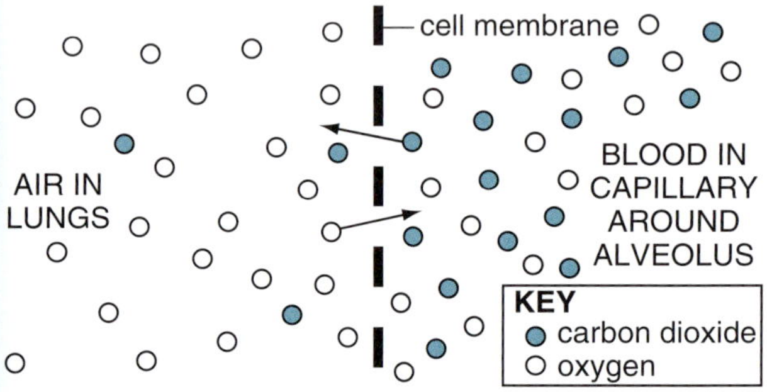

Figure 1.3 Diffusion in the lungs

Dissolved soluble foods and ions can diffuse through cell membranes. This happens in the small intestine where the surface area is increased by villi.

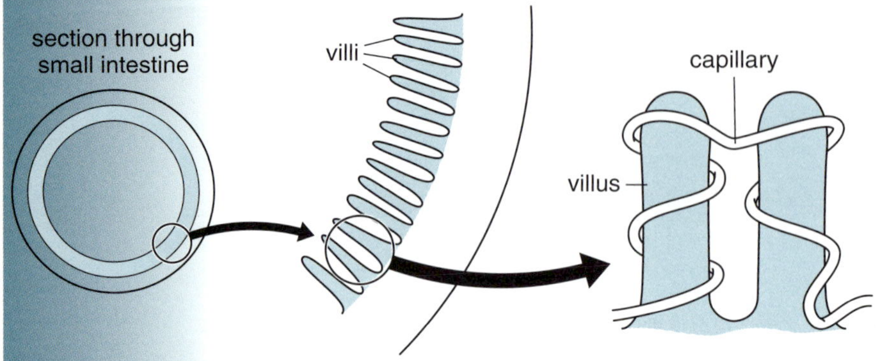

Soluble foods and ions can diffuse quickly into blood as every villus is near to a blood capillary.

Figure 1.4 Diffusion in the small intestine

The digestive system

Enzymes speed up the breakdown of large molecules into smaller molecules. Different enzymes break down different types of food.

Insoluble food	Broken down by enzyme	Soluble food
starch	amylase	sugars
proteins	protease	amino acids
fats	lipase	fatty acids and glycerol

Amylase is made in the salivary glands, pancreas and small intestine.

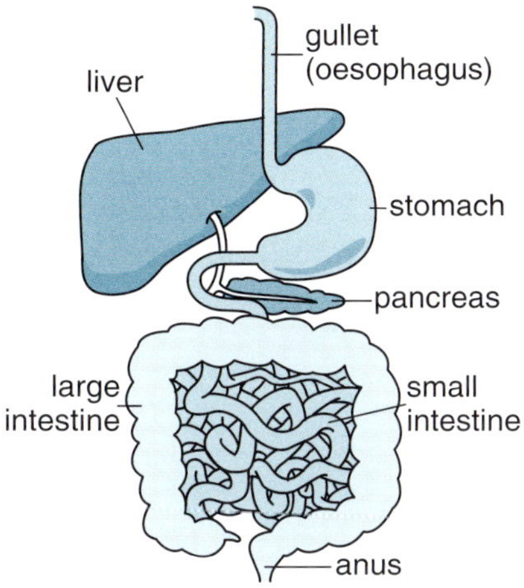

Protease is made in the stomach, pancreas and small intestine.

Lipase is made in the pancreas and small intestine.

Figure 1.5 The main parts of the digestive system

Part of digestive system	What happens here?
mouth	Teeth chew and break up food. Amylase from the salivary glands begins the digestion of starch.
stomach	Produces hydrochloric acid to kill bacteria and make effective conditions for enzymes to work. Protease enzymes begin digestion of protein.
small intestine	Digestion continues. Soluble food is absorbed into bloodstream.
large intestine	Water is absorbed into bloodstream.
anus	Faeces made up of indigestible food leaves at the anus.

The **liver** makes **bile** which is stored in the gall bladder. Bile does three things:

1 It neutralises the acid that was added in the stomach.
2 It makes the right alkaline conditions for the enzymes in the small intestine to work.
3 It breaks fat down into small droplets which have a large surface area.

Summary of digestion

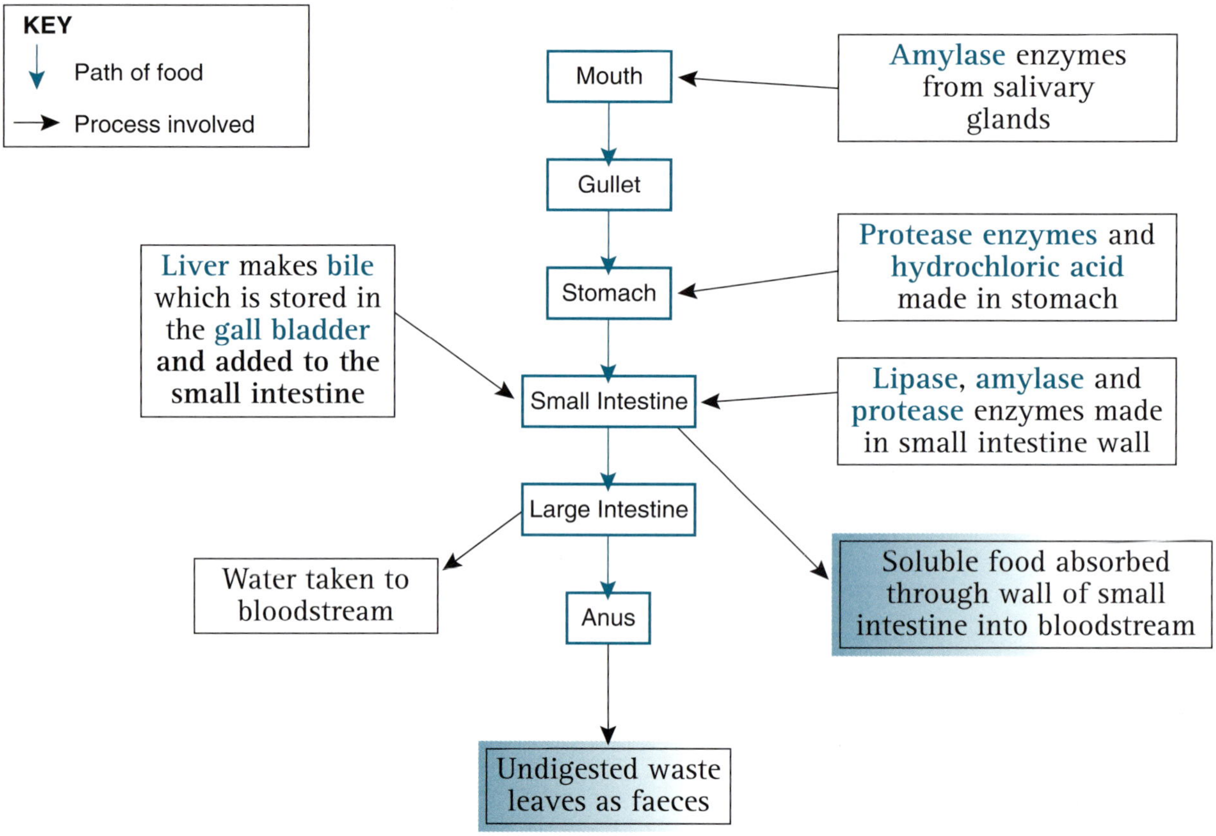

Cover this page and try to redraw the flowchart. Start with the parts of the body, then add the information.

CHECK YOUR UNDERSTANDING

❶ In the lungs, there is always more oxygen in the air than in the blood. Why?

❷ a) What is starch broken down to? b) Which enzyme does this?

❸ a) What is fat broken down to? b) Which enzyme does this?

❹ Name two substances made in the stomach.

❺ Where is soluble food absorbed into the bloodstream?

The breathing system

In the **alveoli** oxygen diffuses into the bloodstream and carbon dioxide diffuses from the bloodstream into the air which is then exhaled.

The **trachea** and **bronchi** are strengthened by rings of cartilage.

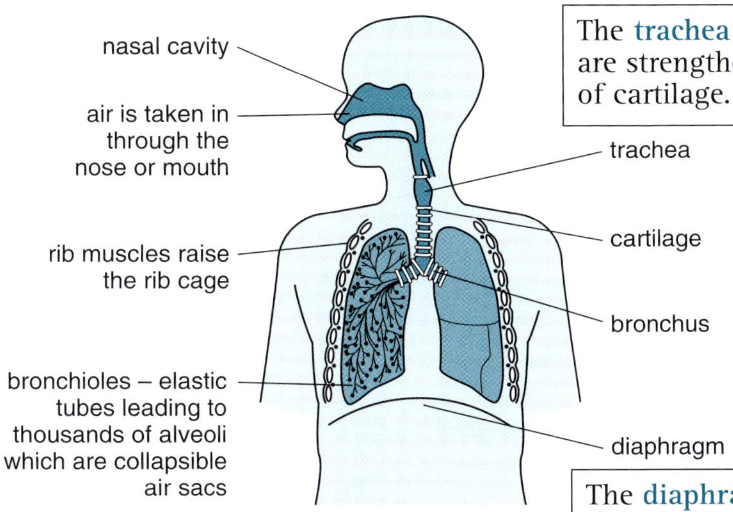

Figure 1.6 The structure of the breathing system in humans

The **diaphragm** is a flexible fibrous sheet with muscles around its margin forming the floor of the chest cavity.

Ventilation – the movement of air into and out of the lungs.

To move air in:	**To move air out:**
The rib cage moves out.	The rib cage moves down.
The diaphragm flattens.	The diaphragm becomes raised.

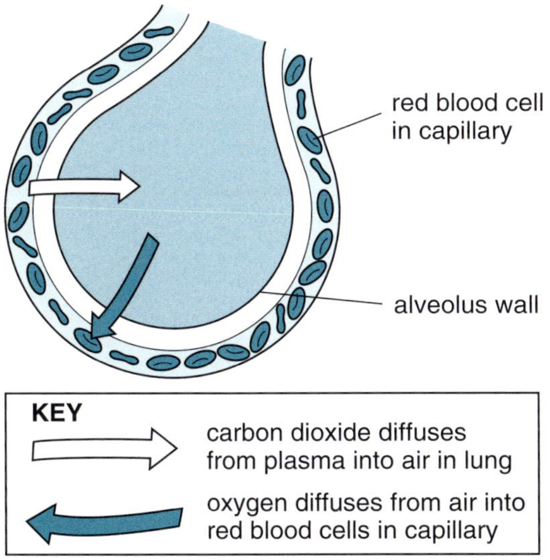

Figure 1.7 An alveolus and its capillary blood supply

Respiration

Respiration happens in all cells. It is a chemical reaction which releases energy from food. The equation below summarises **aerobic respiration**. Aerobic means with oxygen.

glucose + oxygen ⟶ carbon dioxide + water + **energy** transferred

LEARN this word equation.

Aerobic respiration happens when oxygen is plentiful and is happening to you now as you read this book.

If you run quickly across a field, oxygen cannot get to your cells fast enough, so they start to respire in a different way to give you less energy, but still enough to let you run! This type of respiration is called **anaerobic respiration**. Anaerobic means without oxygen.

glucose ⟶ lactic acid + **energy** transferred

Lactic acid is a waste product of anaerobic respiration. This can make your muscles ache and needs to be removed immediately after running. To do this the body takes in extra oxygen (by panting) and the lactic acid is then broken down to carbon dioxide and water. The oxygen that you need after running is called **oxygen debt**, because you 'owe' it to your muscles.

What is the energy used for?

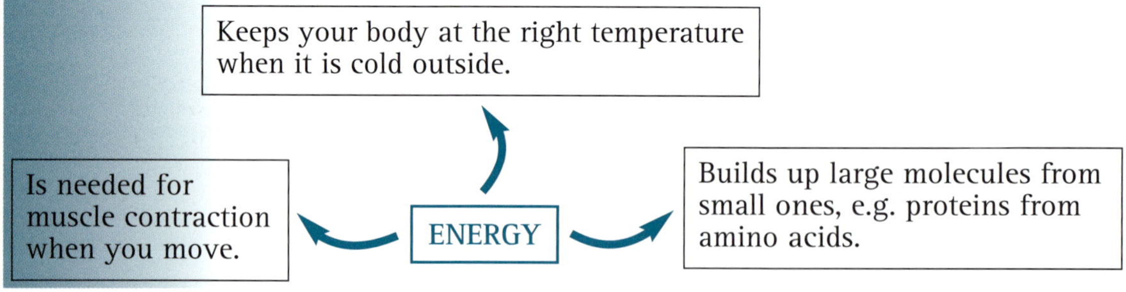

CHECK YOUR UNDERSTANDING

❶ Write the equation for aerobic respiration.

❷ What does anaerobic mean?

❸ What is an oxygen debt?

❹ Name three things that cells use energy for.

The circulatory system

- Useful substances such as food and oxygen must be transported from where they enter the body to all cells.
- Waste products must be transported from the cells to where they leave the body. For example, carbon dioxide is transported from the cells to the lungs.
- The blood (mainly the plasma) transports these substances in the circulatory system.
- The heart is the pump that keeps the system moving.

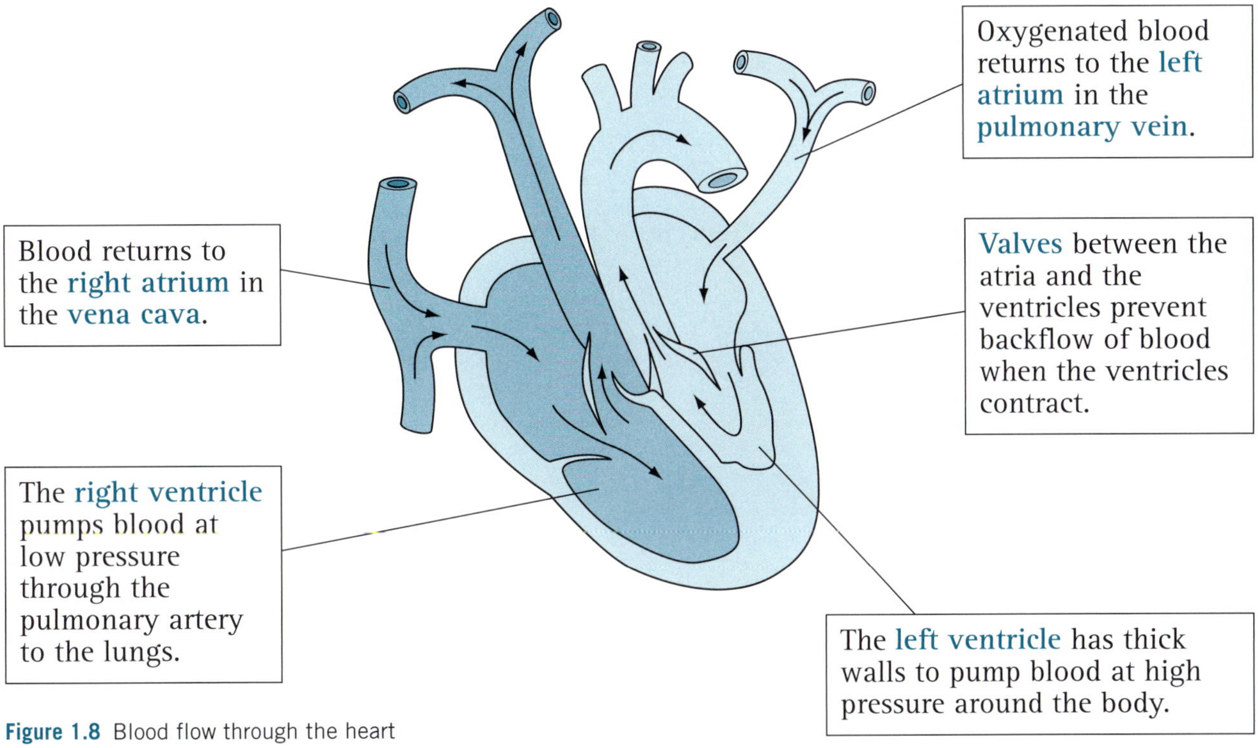

Figure 1.8 Blood flow through the heart

Things to remember about the heart:

- Both sides contract at once.
- The atria (plural of atrium) contract and push blood into the ventricles (gravity helps with this and so the atria have thinner muscle walls than the ventricles).
- The valves between the atria and ventricles close.
- The ventricles contract and push blood in two directions – to the lungs from the right side and to the body from the left side.
- The ventricles then relax and valves in the aorta and the pulmonary artery close.

Blood vessels

Blood leaves the heart in arteries. These have thick walls with muscle and elastic tissue in them. They need to move in and out as blood is pumped through them; this is felt as a pulse.

Veins carry blood back to the heart. They have thinner walls because the blood is no longer at such high pressure. Many large veins like those in the legs have valves to help prevent the backflow of blood as it travels 'uphill' to the heart.

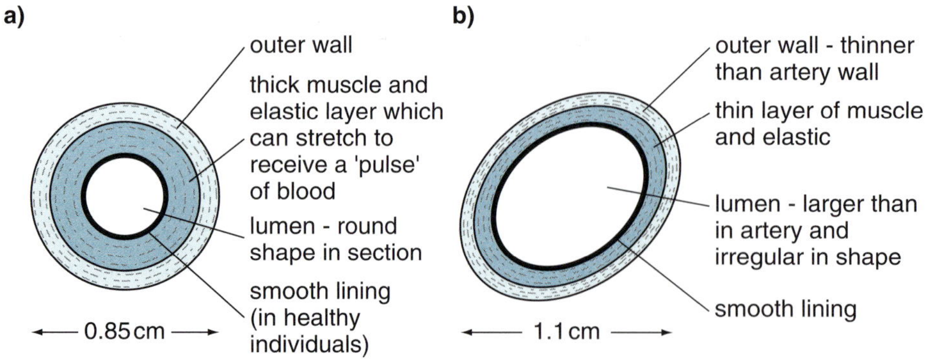

Figure 1.9 Transverse section through a) an artery, b) a vein

Capillaries join arteries to veins. In the muscles and organs, substances pass through the capillary walls into and out of cells. The walls must be very thin (only one cell thick) to allow this to happen.

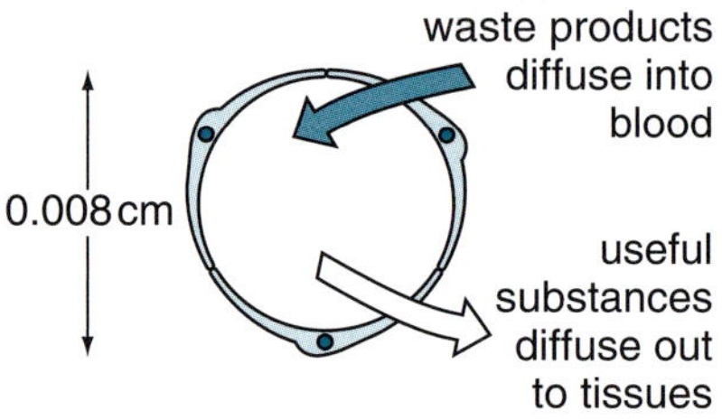

Figure 1.10 Transverse section through a capillary

The capillary wall is a single layer of cells.

BLOOD CONSISTS OF

White blood cells – these help the body to fight disease-causing micro-organisms. They have a nucleus.

Red blood cells – these contain haemoglobin which carries oxygen from the lungs to the cells. They have no nucleus.

Plasma – this is the main transport system of the body. It carries:

- carbon dioxide from the organs and muscles to the lungs
- soluble food from the small intestine to the cells
- urea from the liver to the kidney.

Platelets – these are small pieces of cells without a nucleus. They are involved in blood clotting.

Micro-organisms and disease

Diseases can be caused by micro-organisms called bacteria or viruses entering our body.

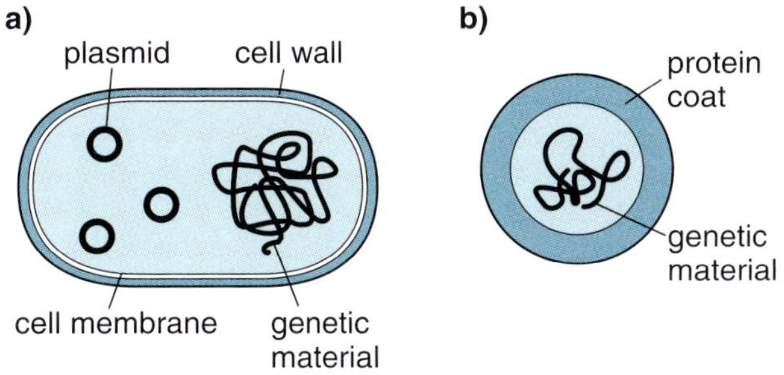

Viruses can only reproduce inside living cells.

Figure 1.11 Section through a) a bacterial cell, b) a virus

To lessen the likelihood of becoming infected with micro-organisms, we should live in hygienic conditions. Some of the things we should do are:

- Wash our bodies and clothes regularly.
- Cook food properly as this kills many bacteria.
- Keep flies from food; they carry bacteria on their bodies.
- Store food in a cool place; bacteria need warmth to grow.
- Stay away from people with infectious diseases; viruses can be passed by coughs and sneezes.

OUR BODY'S DEFENCES

The skin forms a water-tight barrier.

Blood produces clots which seal wounds.

White blood cells kill micro-organisms by engulfing them or producing antibodies to destroy them.

The air passages produce sticky mucus to trap pollen, dust and micro-organisms. This is then swallowed and passes through the digestive system.

White blood cells

White blood cells fight micro-organisms in a number of ways:

- By engulfing and ingesting them.
- By producing antibodies which destroy particular bacteria or viruses. One antibody is needed for each different micro-organism.
- By producing anti-toxins to combat the poisons released by micro-organisms.

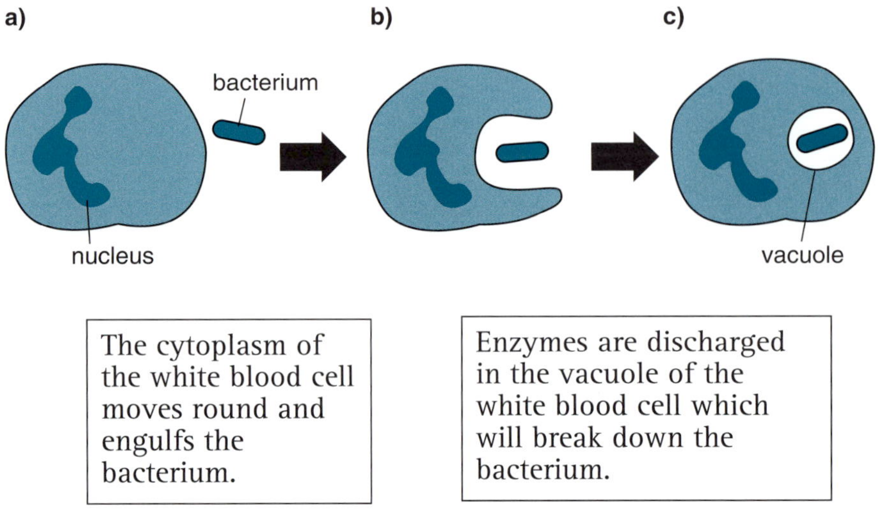

Figure 1.12 A white blood cell engulfing and destroying a bacterium

Vaccination

Babies, young children and people travelling to certain places overseas are given vaccinations by a doctor or nurse. In this process a weakened form of the micro-organism is put into the bloodstream. The white blood cells then make antibodies against the micro-organism. Once the body has made an antibody, it can make it again very quickly. This means that if the real micro-organism enters the body later, the antibodies are ready and waiting to destroy it. The person is now immune to that disease. Sometimes a 'booster' injection is given a few years later to 'top up' the immunity.

CHECK YOUR UNDERSTANDING

❶ Which part of the blood helps to form clots?

❷ List three precautions that we can take to help us avoid getting micro-organisms into our body.

Topic Test

1 Use the information on cells to copy and complete the following sentences.
   a) Sperm cells are adapted for swimming by having a __________. The genetic material is carried in the __________.
   b) Red blood cells have no nucleus so there is room for lots of __________ which carries __________.
   c) Muscle cells are elongated but can __________ when the muscle contracts. (5 marks)

2 Write down the function of these cell parts:
   a) nucleus
   b) cytoplasm
   c) cell membrane. (3 marks)

3 Write a definition for diffusion. (3 marks)

4 Name two ways to increase the rate of diffusion. (2 marks)

5 How is surface area increased in a) the small intestine and b) the lungs? (2 marks)

6 Copy and complete this table to show the digestion of insoluble food to soluble particles.

Insoluble food	Enzyme	Soluble food
starch		sugars
	protease	
		fatty acids and glycerol

(5 marks)

7 Use the words in the box to complete the sentences about bile. Each word should be used once.

neutralises	fat	stomach	surface area	enzymes

Bile __________ acid that was added in the __________.
Bile makes alkaline conditions for the __________ in the small intestine.
Bile breaks down __________ into small droplets so giving them a large __________.
(5 marks)

8 Explain how the ribs and diaphragm allow air to enter the lungs. (2 marks)

9 Write down the sequence of events that happen when the heart contracts. (5 marks)

10 List three things that are carried in the plasma. (3 marks)

11 Explain how a vaccination helps to keep us healthy. (5 marks)

Total: 40 marks

MAINTENANCE OF LIFE

Maintenance in plants

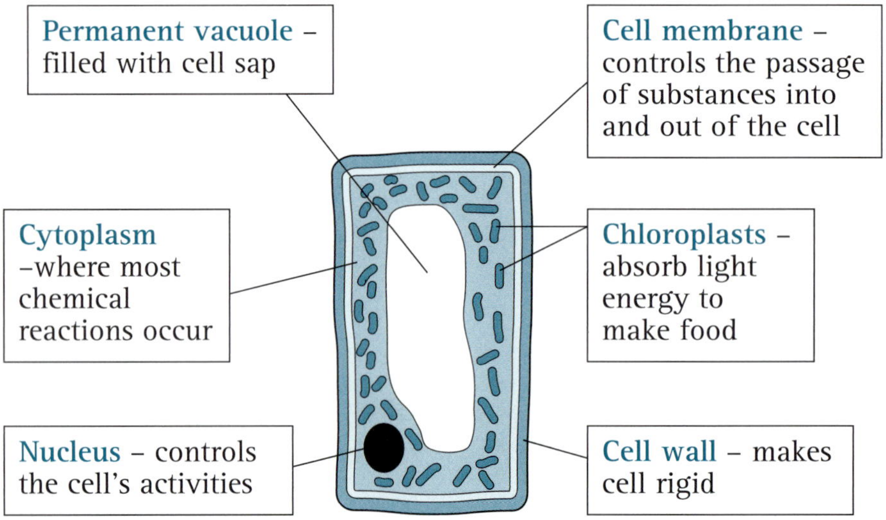

Figure 2.1 A typical plant cell

Different types of plant cell have different shapes which relate to their function.

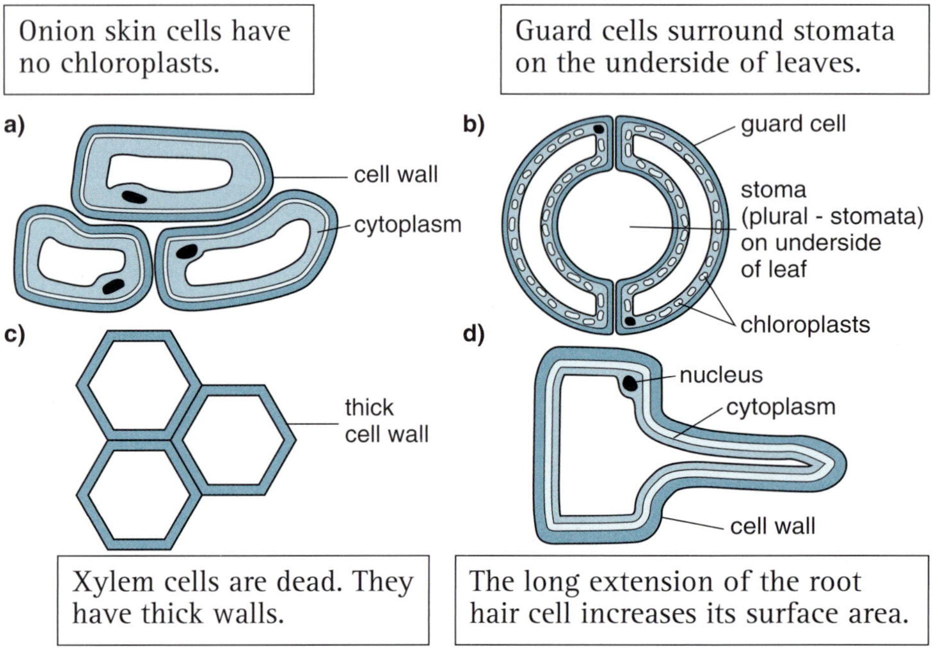

Figure 2.2 Variety in plant cells: a) onion skin cells, b) guard cells, c) xylem cells and d) a root hair cell

Photosynthesis

Photosynthesis is summarised by the equation:

carbon dioxide + water (+ light energy) ⟶ glucose + oxygen

LEARN this word equation.

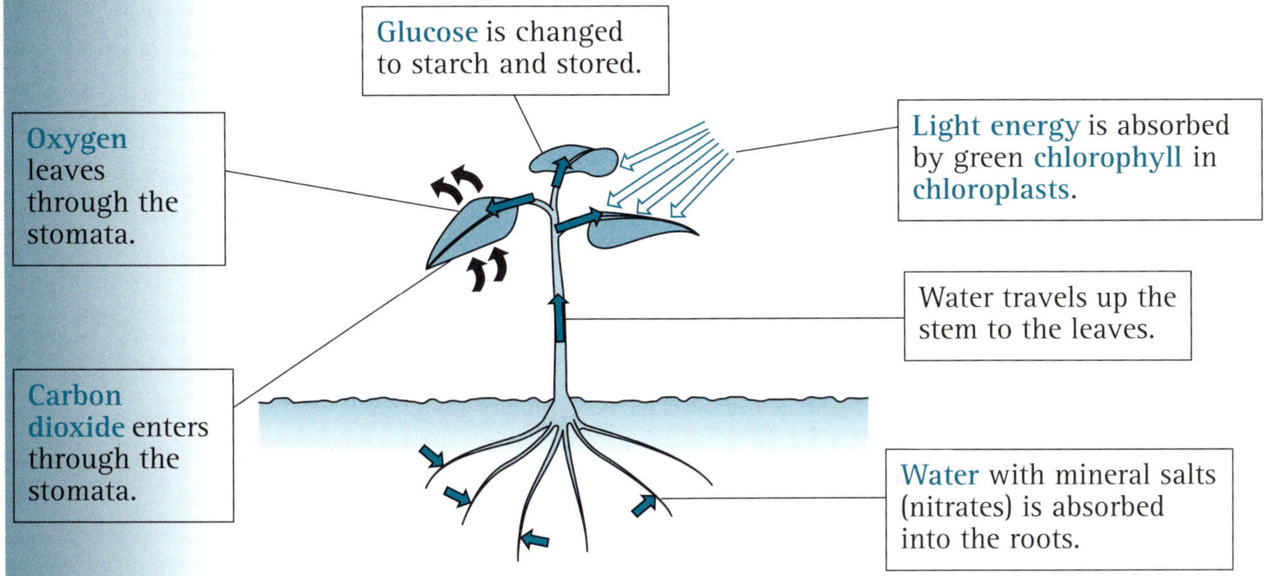

Figure 2.3 How a plant obtains the materials it needs for photosynthesis

The rate of photosynthesis can be reduced by:

- lower temperature
- less carbon dioxide
- less light.

These are called **limiting factors** because they can limit photosynthesis.

The **glucose** made in photosynthesis can be:

- stored as starch
- used as a fuel for respiration by the plant.

Transport in plants

REMEMBER – Diffusion is the movement of molecules from a region of higher concentration to one of lower concentration (see page 2).

Carbon dioxide enters the leaves and the leaf cells by **diffusion**.

Plants lose water vapour from their leaf surfaces by **transpiration**. The rate of transpiration depends on the weather. It is faster in dry, hot, windy conditions. The water evaporates through the **stomata**. These are usually on the underside of the leaf and are controlled by **guard cells**. These cells can change shape to alter the size of the stomata. If a plant loses water faster than the roots replace it, the guard cells close the stomata to reduce water loss. Losing too much water makes the plant wilt. When cells wilt they become floppy or **flaccid**. In young seedlings this means they bend over. When cells are full of water they become firm or **turgid**.

In the **turgid** cell, the cell membrane is pressing against the cell wall.

In the **flaccid** cell, the cell membrane pulls away from the cell wall.

Figure 2.4 What happens in a) a turgid plant cell and b) a flaccid plant cell

Flowering plants have special tissues for carrying water and nutrients.

- **Xylem tissue** carries water up the plant from the roots to the leaves via the stem. (See page 12 for diagram of some xylem cells.)
- **Phloem tissue** carries sugars and other nutrients from the leaves to the rest of the plant. For example in a potato plant, the sugar is carried down to the underground stems we know as potatoes where it is stored as starch. In contrast to xylem, phloem tissue carries material both up and down.

The movement of water is by a special kind of diffusion called **osmosis**.

> **Osmosis** is the diffusion of water from a dilute to a more concentrated solution through a selectively permeable membrane. The membrane allows water molecules to pass through but not solute molecules.

LEARN this definition.

Figure 2.5 shows how water enters the plant from the soil. The root hair cells increase the surface area of the roots so that they can take in the water more quickly.

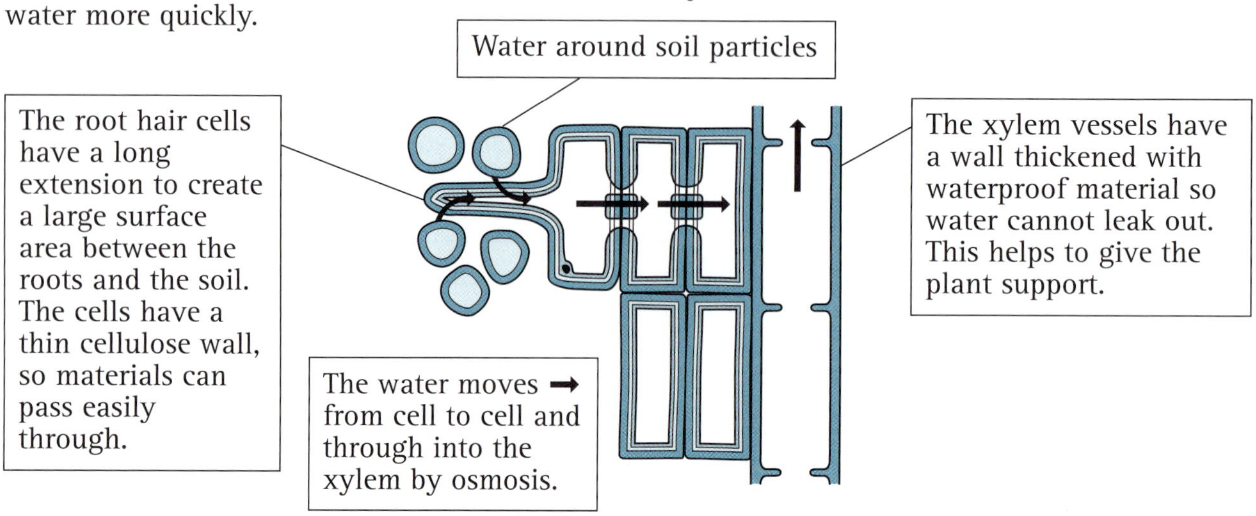

Figure 2.5 The role of the root hair cell in the uptake of water and the movement of water through the root to the xylem

CHECK YOUR UNDERSTANDING

❶ What sort of weather conditions are needed for a plant to transpire rapidly?

❷ Plants have a waxy cuticle on their leaves to stop too much water loss through them. Why do plants which live in deserts have a thicker cuticle?

How plants respond to their surroundings

Plants are sensitive to the direction of:

- gravity
- water
- light intensity.

As a result their roots grow downwards where the water is and their shoots grow upwards towards the light.

Response of plant shoots to light

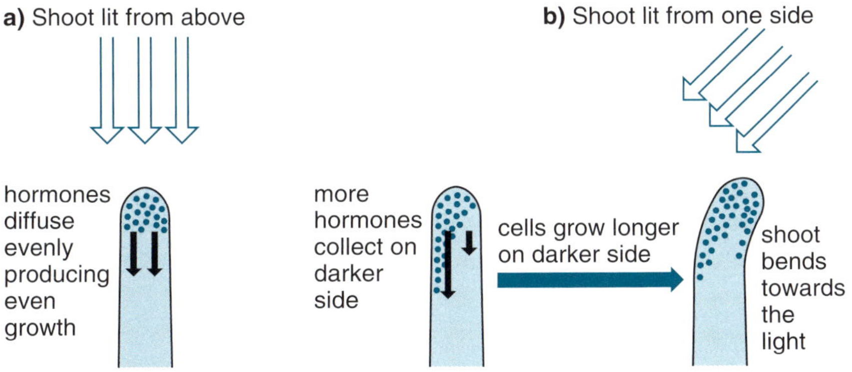

Figure 2.6 Plant responses to light

Plant shoots grow towards the light due to the uneven distribution of **hormones** which collect on the side away from the light and make the plant cells there grow faster.

Response of plant roots to gravity

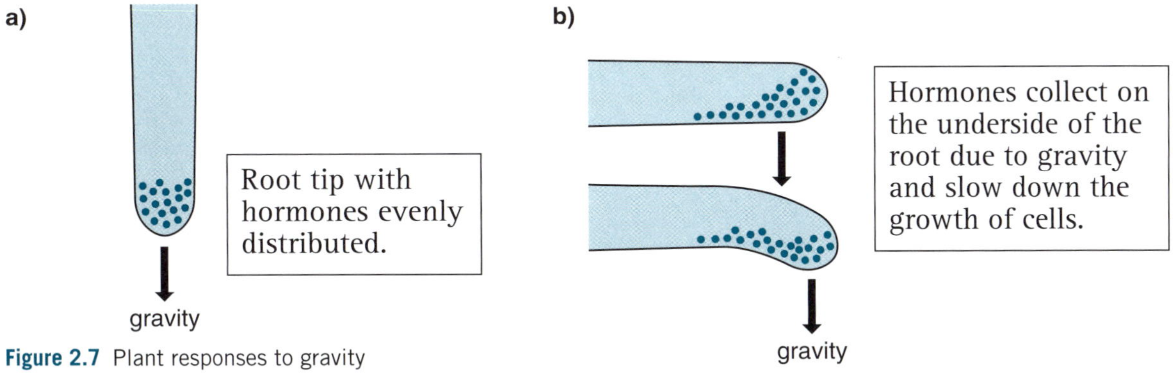

Figure 2.7 Plant responses to gravity

The hormones collect in the bottom of the root and so the root grows downwards. If the root is placed on its side the hormones collect at the base, and again gravity makes the root grow downwards.

Plant growers can use plant hormones to:

- Stimulate the growth of roots from cuttings to get lots of identical plants.
- Kill weeds by making them grow fast, get weak and die.
- Control the ripening of fruit on the plant. This allows it to ripen more slowly so that it can be transported from one country to another and still arrive in good condition.

Maintenance in humans

How humans respond to their surroundings

The nervous system allows us to react to our environment and co-ordinate our behaviour. Cells called receptors detect stimuli. The table shows where the receptors are and what they do.

Receptors	Function
in the eye	sensitive to light
in the ears	sensitive to sound
in the ears	allow us to keep our balance
on the tongue	sensitive to chemicals; enable us to taste
in the skin	sensitive to touch, pressure and temperature changes

The eye

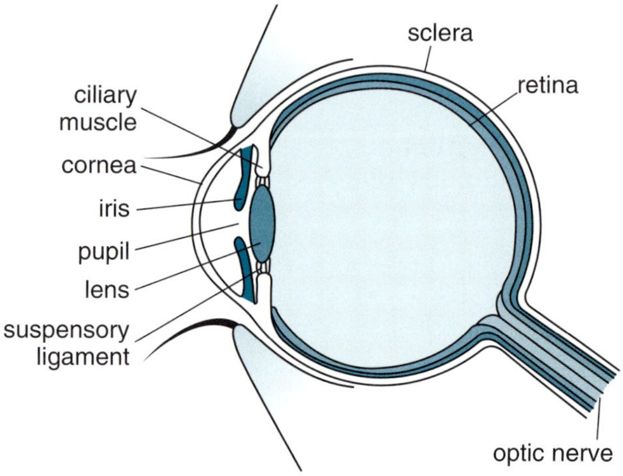

Figure 2.8 A vertical section through a human eye

Parts of the eye:

- The sclera. A tough outer layer which is transparent at the front where it is called the cornea. The cornea bends the light when it enters the eye.

The receptor which you need to know in some detail is the eye. You should be able to label the following parts and know their different functions.

- The **iris**. This is muscular and controls the amount of light entering the eye by controlling the size of the **pupil**.
- The **lens**. This is held in position by **suspensory ligaments** and **ciliary muscles**. The lens focuses the light on to the retina.
- The **retina**. This contains light-sensitive receptor cells. These send impulses to the brain along sensory neurones in the **optic nerve**.

Reflex actions

Information from the receptors passes along **neurones** in nerves to the brain. The brain then co-ordinates the response. Sometimes these responses are rapid and automatic and are called **reflex actions**. We don't think about them, and they are often protective.

Examples of reflex actions are:

- Blinking when an object comes towards the eye – to protect the eye from damage.
- The widening of the iris to narrow the pupil when a bright light is shone in the eye – to protect the retina from too much light.
- Removing the hand from a hot object – to protect the skin from burning.

A footballer saving a goal is not a reflex action as he has to think about it and decide what to do!

In a simple reflex action, **electrical impulses** pass from the receptor along a **sensory neurone** to the **spinal cord** or brain. They then pass from the spinal cord along a **motor neurone** to a muscle or a gland. The muscle or gland then brings about the **response**.

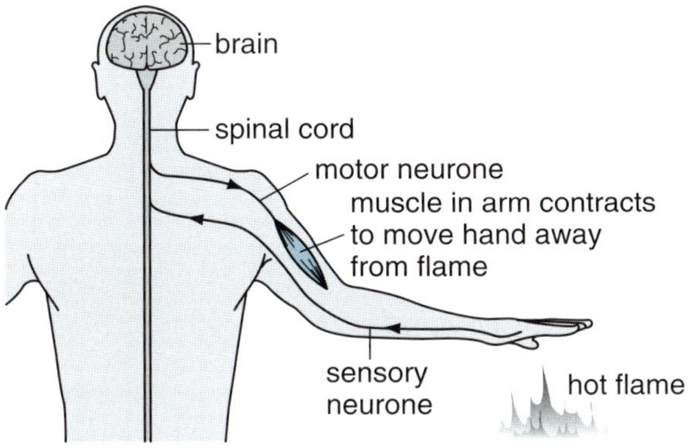

Figure 2.9 A reflex action

CHECK YOUR UNDERSTANDING

1. Name the receptors sensitive to light, sound and taste.
2. Why is saving a goal not a reflex action?
3. Give two examples of reflex actions in our body.

How the body maintains its internal conditions

Automatic systems in the body ensure that:

- Body temperature is correct.
- The right amount of water and sugar is in the bloodstream.

Removing waste products

- **Carbon dioxide** is produced by all cells when they respire. It is taken to the lungs in the bloodstream and then breathed out.
- **Urea** is produced by the liver when extra amino acids not needed by the body are broken down. The urea is taken to the **kidney** by the bloodstream and dissolved in water to form urine. It is temporarily stored in the bladder.

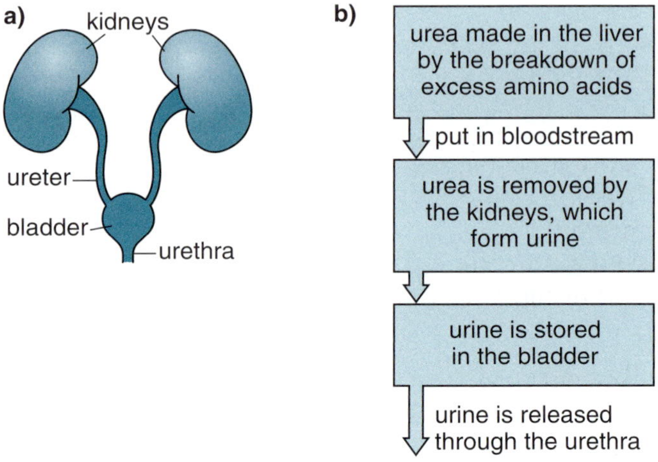

Figure 2.10 a) The excretory system in humans. b) The process of removal of urea

Controlled internal conditions include:

- The **water content** of the body. Water leaves the body via the lungs when we breathe out, via the skin when we sweat and via the kidneys when we pass urine.
- The **ion content** of the body (sometimes called mineral salts). Ions are lost through the skin when we sweat and through the kidneys in urine.
- The **temperature** must be controlled so that **enzymes** can work at their best. Sweating helps to cool the body by evaporation. When it is hot we drink more to balance the water lost in sweat.

The sugar content of the blood is controlled by **hormones**. These are proteins made in glands and carried in the bloodstream to their target organs.

If blood sugar is too high, **insulin** causes the liver to store some as **glycogen**. If blood sugar is too low, **glucagon** causes the liver to change some glycogen back to sugar.

Hormone	Gland producing hormone	Target organ
insulin	pancreas	liver
glucagon	pancreas	liver

A **diabetic** person does not make enough insulin and has to either inject insulin or control the sugar content of the food eaten.

CHECK YOUR UNDERSTANDING

❶ Name two waste products that must be removed from the body.

❷ Name the hormone which lowers the blood sugar level.

The effect of drugs

A **drug** is a chemical that changes the way our bodies work or changes the way we think or react. Even drugs which are used legally may harm us, so we need to know the dangers involved.

If we use drugs a lot, they change the chemical processes in our bodies so that we become dependent on them. This means we feel very ill if we don't have the drug. Sometimes we can become **addicted** to the drugs. This means our bodies will show physical symptoms (withdrawal symptoms) if the drug is taken away. These symptoms can include vomiting, headaches, sweating and stomach pains.

Solvents

Young people sometimes experiment with **solvents** such as glue or lighter fuel.

Solvents can cause many problems.

- They cause the person to behave in strange ways.
- They may cause permanent damage to the lungs, the liver and the brain.
- If sniffed from a polythene bag, the person may suffocate because the bag stops them breathing.

Alcohol

Another widely used but legal drug is **alcohol**. Alcohol is found in drinks such as beer, wines, spirits and 'alcopops'.

If alcohol is taken in large quantities it can:

- Affect the nervous system by slowing down reactions. This is especially dangerous if the person then drives a car. Even walking down the road can be dangerous.
- Lead to lack of self control; some people are more likely to fight when they have had too much alcohol to drink.
- Lead to unconsciousness or even a coma.
- Damage the liver and the brain.

CHECK YOUR UNDERSTANDING

1. What is a drug?
2. Give two examples of drugs.
3. What damage can alcohol do to someone who drinks too much over a long period of time?

Tobacco

The addictive substance in tobacco is **nicotine** and after just a few cigarettes, some people find it very difficult to stop the habit.

The effects of smoking

The airways of the breathing system are lined with cells which have tiny beating **cilia** on them. These cilia set up a wave motion that takes the mucus made by other cells up to the throat, where it is swallowed. The mucus traps dust, pollen and carbon particles and stops them from reaching the lungs. Tobacco smoke can stop the cilia from working so that the mucus builds up and falls into the lungs. The person then has to cough to remove it. Severe coughing can damage the lungs.

Emphysema is an illness where the tiny alveoli are broken and so the surface area of the lung is greatly reduced. People with emphysema are always short of breath due to lack of oxygen.

The smoke contains substances like **tar** which can cause **lung cancer**. Smoking cigarettes can also cause diseases of the heart and blood vessels.

Some facts about cigarette smoking

NICOTINE:
- is addictive – making it difficult to stop smoking once you've started
- makes blood clotting more likely
- makes the smoker feel more relaxed
- makes boring jobs more tolerable.

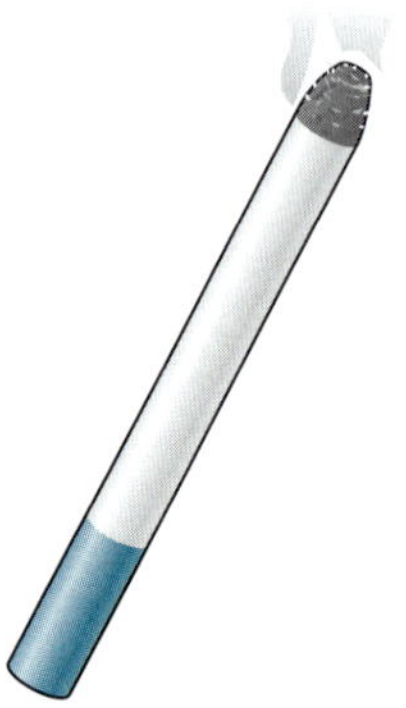

CARBON MONOXIDE:
- combines with haemoglobin so less oxygen is carried by the blood. In pregnant women this means that the fetus gets less oxygen and is often small when born.

TARS:
- are deposited in lungs
- cause bronchitis
- cause lung cancer
- cause extra mucus secretion.

Smoking and cancer

A link between smoking and ill health was first noticed by the ship's doctor who travelled with Columbus to South America. But no scientific studies were carried out until the late 1940s, when three studies all showed the same link. Since then, all investigations have shown that smoking greatly increases the risk of lung cancer.

CHECK YOUR UNDERSTANDING

❶ Name two ways in which the cells lining the breathing tubes of a healthy person help to keep the lungs free from dust and pollen.

❷ Name three diseases caused by smoking.

Topic Test

1 Write down the function of a) the cell membrane, b) the cell wall, c) the chloroplasts. (3 marks)

2 How does the shape of the root hair cell relate to its function? (1 mark)

3 Copy and complete the following sentences.

Light is absorbed by __________ in chloroplasts. Glucose is changed to __________ and stored in the __________. Carbon dioxide and oxygen travel through the __________ to enter and leave the leaf. (4 marks)

4 Write the word equation for photosynthesis. (1 mark)

5 What are the limiting factors for photosynthesis? (3 marks)

6 Copy and complete these sentences.

Water travels up the plant in the __________ tissue. The phloem tissue carries __________ up and down the plant. (2 marks)

7 Give two ways that commercial growers of plants use hormones. (2 marks)

8 Fill in the gaps in the table.

Receptors	What it detects
	touch and pressure changes
tongue	

(2 marks)

9 a) What does the eye lens do?
b) What does the iris do? (2 marks)

10 Name three internal conditions that must be controlled by the body. (3 marks)

11 What does insulin do? (1 mark)

12 What does 'being dependent on a drug' mean? (1 mark)

Total: 25 marks

ENVIRONMENT

Why do plants and animals live where they do?

Plants and animals are said to be adapted when they are able to survive in their normal environment. The conditions they need to grow and reproduce are available. Some animals have had to adapt to extremely harsh environments. Some examples are given in the table below.

ARCTIC ENVIRONMENT		DESERT ENVIRONMENT	
Arctic conditions	**Adaptations**	**Desert conditions**	**Adaptations**
Very cold	Thick fur to trap a thick layer of air. Air is a poor conductor of heat. Thick layer of fat (or blubber in whales). Fat is a good insulator.	Very hot	Thin fur on top of the body. Little body fat so heat is easily lost from skin capillaries.
Very cold	Compact body shape with small ears to reduce surface area for heat loss. Small feet to reduce heat loss by conduction.	Very hot	Body shape gives maximum surface area for heat loss. For example, Fennec foxes have large ears to radiate heat and long, thin legs.
White background in winter	White fur in winter for camouflage and to reduce radiation heat loss (e.g. Arctic fox).	Sandy background	Fur of sandy brown colour for camouflage.

Adaptations of plants to arid conditions

Arid or very dry conditions are found in deserts where it may not rain for several years and there are sand dunes which are rapid draining.

Cacti are well adapted to desert conditions.

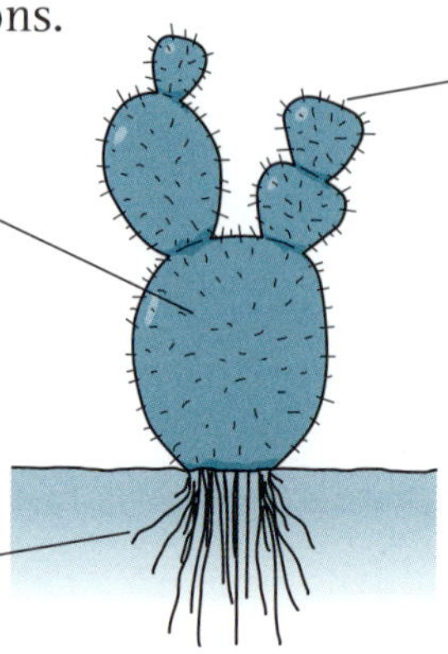

Figure 3.1 Adaptations of a cactus

Plant and animal populations

The population of plants in a community will be affected by:

Competition for light – trees may shade smaller plants so that they do not get enough light for photosynthesis.

Grazing e.g. by caterpillars may reduce the number of leaves so less photosynthesis takes place and so less growth.

Competition for space to grow.

Competition for mineral ions such as nitrates which are needed for making proteins.

Competition for water from the soil – water is needed for turgor support.

Figure 3.2 Factors affecting plant population

Count the number of times the idea of competition has been used on this page. The plants and animals really do affect each other and can change the environment.

The population of animals in a community will be affected by:

- The **amount of food** available and the **number of animals competing** for the same food source. For example, if there are too many grazing animals, the grass will be eaten and trampled to ground level and will not grow so the population of herbivores (sheep or rabbits) will be reduced.
- The **proportions of predators and prey**. If there is more food for the prey, they will rear more young providing more food for the predators so they will also breed more. When the number of predators (e.g. foxes) increases, they eat more prey (e.g. rabbits), so the number of prey will soon be reduced.
- **Disease** – illnesses spread easily if the population gets too large and close packed.

Predator – An animal which kills and eats other animals.

Prey – Animals that are eaten by another animal.

CHECK YOUR UNDERSTANDING

1. Explain how polar bears are adapted for surviving in very cold conditions.
2. What do animals compete for in their environment?
3. Why are there few plants on a woodland floor?
4. The disease myxomatosis killed many rabbits in the UK. Suggest how this affected a) plant growth and b) the fox population.
5. If many of the leaves on an apple tree were eaten by caterpillars in the spring, why would the apple crop be very much reduced that autumn?

Energy and biomass in a food chain

Make sure you use the words sunlight energy or light energy NOT Sun. It is the ENERGY which is important.

Green leaves capture a small percentage of **light energy**.

Plants use some energy for making sugars, cellulose, and proteins – i.e. making plant **biomass**.

Some of the energy in the plants eaten by a **herbivore** is used for **animal growth** – i.e. stored as more protein in the muscles.

Some energy is used for **plant respiration** and is therefore lost from the food chain.

Some of the energy of the herbivore eaten by the **carnivore** (blue tit) is made into **muscle protein** as the carnivore grows.

At each link in the food chain, some of the energy is used by the animal for **respiration** – that is making chemical energy for use in the cells, use by muscles and for keeping the body warm. Some energy is lost in urine and faeces.
Energy is lost from the food chain and there is less to be passed on at each link.

Carnivores may be eaten by bigger carnivores. More of the energy is converted into **biomass**.

Figure 3.3 What happens to energy and biomass in a food chain

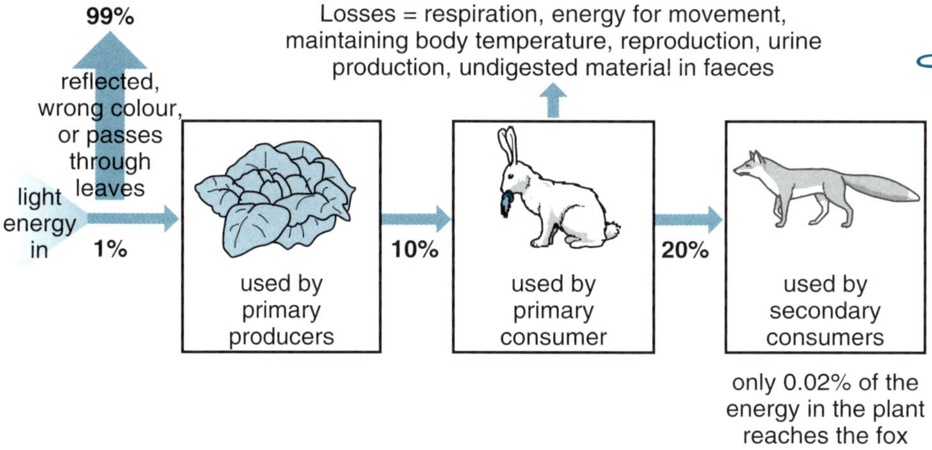

Figure 3.4 A simple three-link food chain showing where the energy is lost in the chain

If you are given numbers, work out what they tell you – they won't be difficult. In Figure 3.4, take the energy from the Sun as equivalent to £100. Work out the amount the lettuce traps: 1% of £100 = £1. The rabbit gets 10% of £1 which is 10p. So how much does the fox get? This shows that you can't have a food chain with too many links as there will be very little energy for the last carnivore.

THINK – From your knowledge of the organisms in this food chain, what changes in size would you see as you pass along the chain and what changes in population numbers in a given area?

Applying this idea to a food web, we see that at each feeding level the animals are larger but there are fewer of them. The biomass (mass of living material) at each level gets less – this can be drawn as a pyramid of biomass.

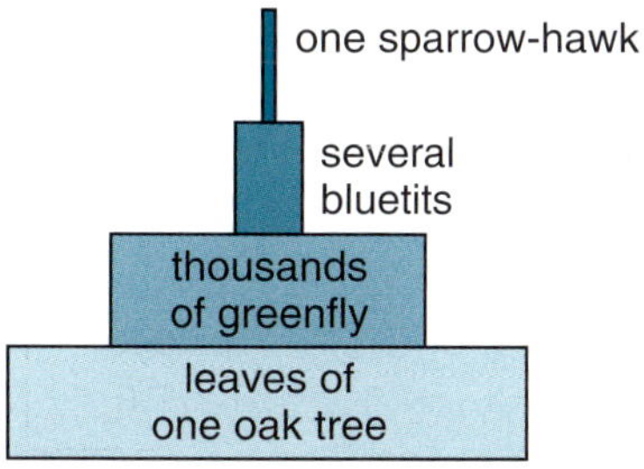

Figure 3.5 A pyramid of biomass for an oak tree

Make sure that your rectangles get smaller as the biomass reduces.

CHECK YOUR UNDERSTANDING

1. Give one reason why green plants trap so little of the sunlight energy which falls on them.
2. Complete the following statements:
   a) Mammals use a lot of energy for respiration as they keep a high body ___________ .
   b) Animals which run around a lot use energy for ___________ .
3. For human food production, why would a three-link chain be more efficient than a five-link chain?
4. What is wrong with this statement: A pyramid of biomass shows that most biomass is stored in the herbivores.

Nutrient cycles

Plants take up mineral ions (nitrates etc.) from the soil and use them to make proteins. → Animals eat plants and make animal protein. → Animals produce urine, faeces and then die and plants drop their leaves: all are returned to the soil. → Micro-organisms break down the waste and dead material and change it back to mineral ions.

Microbes or micro-organisms are bacteria and fungi which release enzymes to digest the waste and dead material. They can't use it all for growth so substances returned to the soil are taken up by plant roots and the cycle starts again. Micro-organisms are most active in warm, damp conditions with lots of oxygen.

Man makes use of the decomposition of waste by micro-organisms in sewage works and for composting. A lot of household waste can now be composted.

LEARN the cycle and the conditions in which micro-organisms break down dead material fastest.

Carbon

- The element carbon is very important – it is found in a large number of biological compounds such as sugars, starch, fats, proteins and cellulose. These compounds make up plant and animal biomass.
- Carbon is in the gas carbon dioxide produced by respiration.
- Carbon is constantly being cycled in the environment.

The carbon cycle

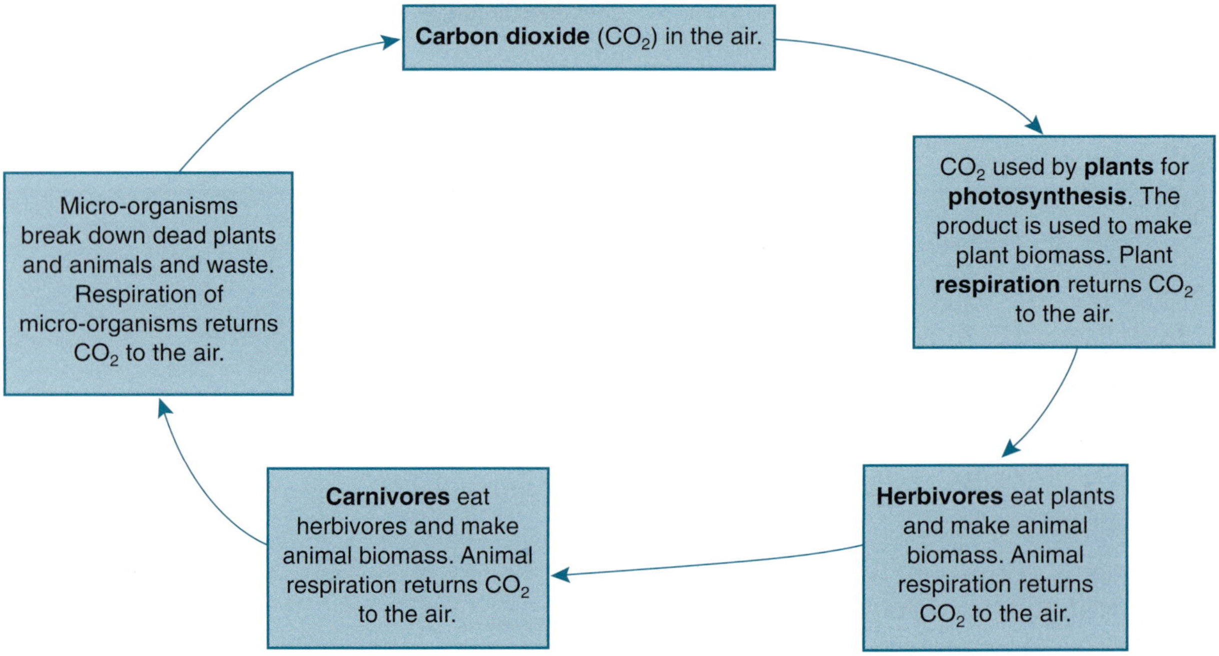

CHECK YOUR UNDERSTANDING

❶ In a stable community the 'take up' and 'return' are balanced, but what happens if man removes or harvests plants from the environment?

❷ What are the ideal conditions for composting?

❸ What is the only process that removes carbon dioxide from the air?

How do humans affect the carbon dioxide balance?

Humans have increased the carbon dioxide levels by:

1 **Deforestation** – cutting down and burning trees to use the land for cattle ranching.

2 **Farming** – this increases the number of bacteria and fungi in the soil and thus the amount of respiration.

3 **Transport** – burning **fossil fuels**, for example petrol and diesel.

4 **Generating electricity** – burning coal, oil and gas in power stations.

Deforestation also means that less carbon dioxide is removed from the air – fewer trees so less photosynthesis. This means that less carbon dioxide is locked up in the cellulose of wood.

The effect of humans on the environment

Rapid growth in the human population has increased demand for:

1 Land for building.

2 Building materials – leading to increased quarrying.

3 Energy – so more power stations are needed with the result that non-renewable fuel resources (coal, oil and natural gas) are rapidly being used up.

4 Food – resulting in more intensive farming and thus more pesticides and fertilisers being used.

5 Transport of materials from producers to towns.

More people produce:

1 More sewage which needs treatment to prevent sea or river pollution.

2 More refuse – we are rapidly running out of landfill sites so waste must be recycled, composted or burnt.

3 Air pollution from industries and transport.

4 Water pollution if toxic chemicals from industry or excess fertilisers or pesticides from intensive farming wash into rivers.

Acid rain

Burning fossil fuels releases acidic gases like sulphur dioxide and nitrogen oxides into the air.

The acidic gases dissolve in the water in the air and fall as acid rain (rain with a low pH).

Acid rain can directly affect plant leaves, resulting in less photosynthesis and less growth.

When acid rain falls in lakes, it reduces the pH of the water, killing plants, insects and fish.

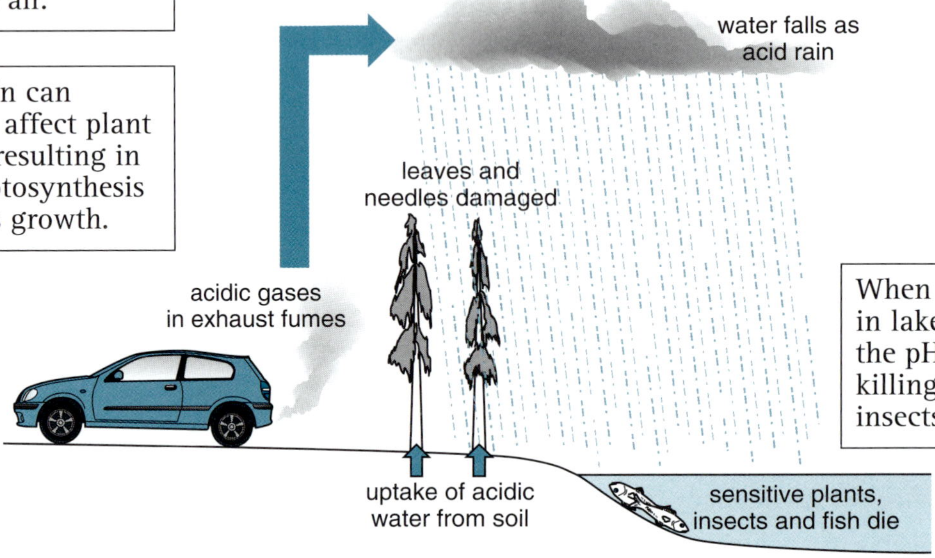

Figure 3.6 The effects of acid rain

Ways to reduce the acid rain problem

- Use of low sulphur fuels.
- Use of bulk transport for goods – railways.
- Less use of private cars – better public transport.
- Use of alternative renewable energy sources like wind power, solar power or hydroelectric power.

Global warming

Atmosphere
Levels of carbon dioxide and methane are rising hence increasing the 'greenhouse effect'.

More cows produce more carbon dioxide from respiration, and more methane from the fermentation of grass.

More rice paddy fields – the bacteria in the flooded fields release methane.

More combustion of fossil fuels produces more carbon dioxide.

The result of the 'greenhouse effect' is global warming. The change in temperature will change climates resulting in more storms and more droughts. The melting of the polar ice caps and of glaciers will cause flooding in low coastal areas such as Bangladesh, Holland and parts of eastern England.

Ways to reduce the problem of global warming

- Develop alternative energy resources so we burn less fossil fuels.
- Increase use of public transport for passengers and bulk rail transport for goods.
- Eat more vegetable protein and less beef.
- Increase energy efficiency in homes with better insulation and low-energy light bulbs.

CHECK YOUR UNDERSTANDING

1. Write a bullet point sequence for the causes of acid rain.
2. Copy and complete this table.

Gas	Carbon dioxide	Methane	b)	Nitrogen oxides
Excess causes	Global warming	a)	Acid rain	c)

LEARN the effect that each gas has on the environment.

3. Landfill sites are running out. What other ways are there of getting rid of domestic waste?
4. Give three ways in which water can be polluted.
5. What problems would be reduced if people used their cars less?
6. In recent years England has suffered dry summers and winter storms. What is the suggested cause of these climate changes?

Topic Test

1 Which of the following measures will help to reduce the greenhouse effect?

a) Use fewer aerosol sprays.
b) Use public transport rather than large number of cars.
c) Less insulation in buildings.
d) Change to high-efficiency light bulbs.
e) Eat more vegetable protein and less beef.
f) Plant more trees.
g) Increase the area of paddy fields.
h) Develop the use of wind farms to generate electricity. (5 marks)

2 Match each adaptation to the reason for that adaptation.

Adaptation	Reason for adaptation
a) Cacti are adapted to the desert as they have thick fleshy stems	i) for maximum surface area for heat loss
b) Desert foxes have large ears	ii) as they contain chlorophyll for photosynthesis
c) Leaves are reduced and form spines	iii) to collect as much water as possible after rainfall
d) Cacti roots are spread over a wide area	iv) as this is a good insulator
e) Whales have a thick layer of fatty blubber	v) which store water
f) Cacti stems are green	vi) to reduce transpiration and stop the plant being eaten

(6 marks)

3 Read the following and then answer the questions. Use the correct terms to explain your answer.

Every year two pairs of robins nested in the orchard and each pair reared four young and fed them on greenfly and small caterpillars. In the spring following a mild winter another two pairs moved in. In that year each pair only raised two young.

a) Suggest a reason for the reduction in the number of chicks raised.

The following year the orchard was visited regularly by a sparrowhawk.

b) Suggest what might happen to the robin population.

c) From the information above, draw a food web for the orchard.

(6 marks)

4 Composting is taken very seriously by some gardeners. Fred regularly turns his compost heap to make sure that it is not too compacted, he checks that it is moist but not waterlogged and has it covered to keep it warm.

Copy and complete the following statements:

a) He turns the heap to ________________________________ .

b) It is kept moist to ________________________________ .

c) It is covered to keep it warm because ____________________ .

(5 marks)

5 a) Copy and complete the food chain by writing above the arrows the percentage of energy passed on at each stage in the food chain.

SUN ⟶ PLANT ⟶ HERBIVORE ⟶ CARNIVORE

b) Why is energy lost between the herbivore and the carnivore?

c) What changes could a pig farmer make to increase the level of meat productivity in his pigs?

(8 marks)

Total: 30 marks

INHERITANCE AND SELECTION

Types of reproduction

There are two methods of reproduction:

1 Asexual reproduction

Asexual reproduction involves only *one* parent. Offspring or new cells are formed by normal cell division, so they are genetically identical.

This can be seen in plants which have runners.

REMEMBER – Individuals in a clone have no genetic variation.

2 Sexual reproduction

Sexual reproduction requires *two* parents and the joining of sex cells produced by the two parents to form the offspring.

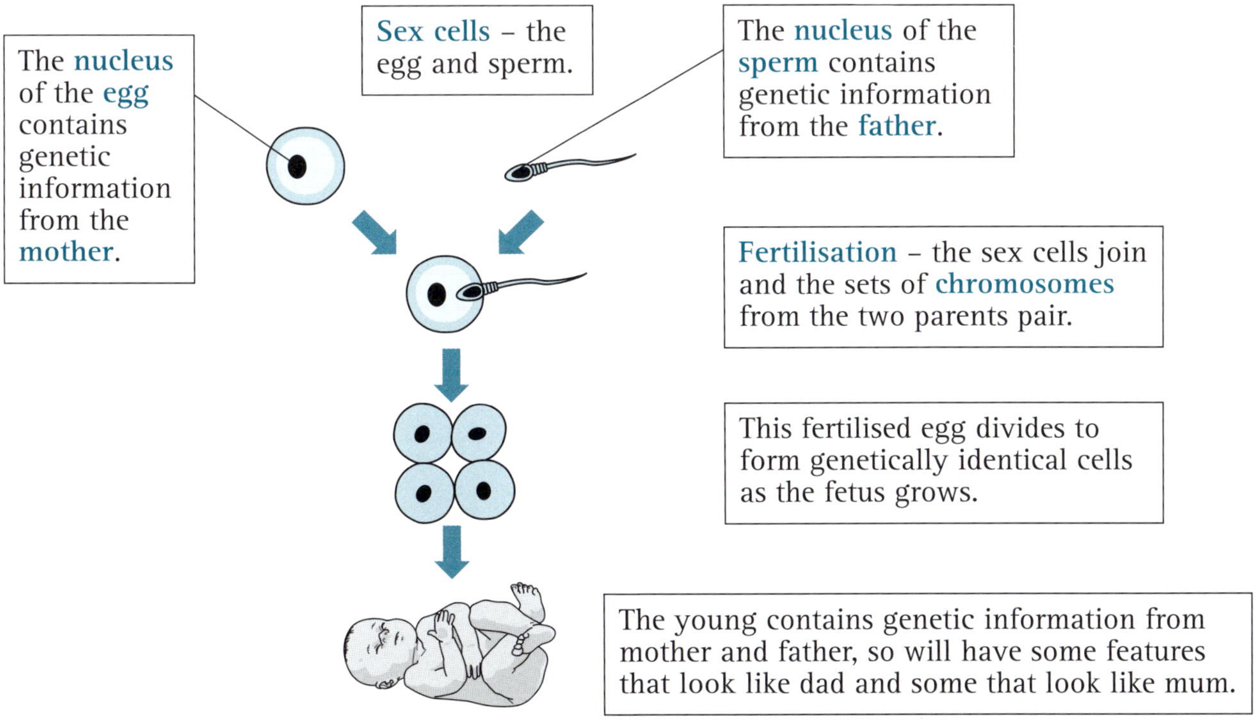

Figure 4.1 The process of sexual reproduction

Environmental factors, e.g. light, nutrients and water, can cause differences in the appearances of plants and animals produced by sexual or asexual reproduction.

Variation within a species

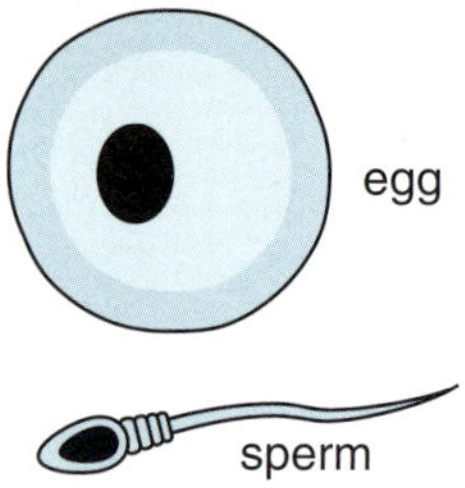

Figure 4.2 Sex cells

The nucleus of a sex cell contains one set of chromosomes.

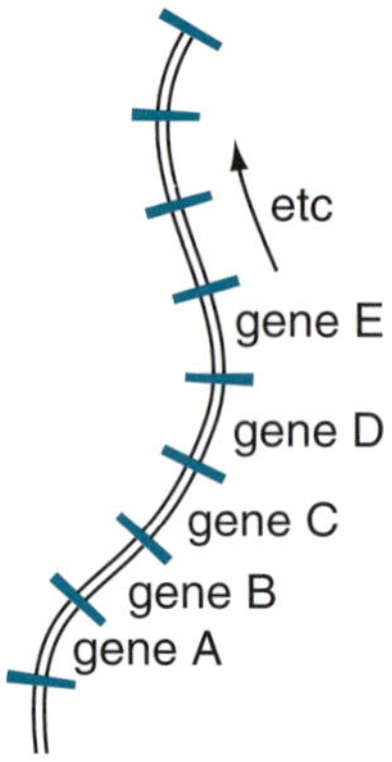

Figure 4.3 A chromosome

Each chromosome carries a line of many genes.

Different genes control the development of different characteristics.

There are different forms of some genes and these are called alleles, e.g. the eye colour gene has alleles for blue eye colour, brown eye colour and green eye colour.

REMEMBER – Genetic variation comes from sexual reproduction. No two offspring are alike unless they are identical twins.

In sexual reproduction the young receive a random selection of alleles from both parents so each individual has a unique set. When the chromosomes with their alleles from mother and father pair, only the dominant allele shows in the offspring, so the offspring will show some features from dad and some from mum. (More detail on pages 37–38.)

CHECK YOUR UNDERSTANDING

1. Copy and complete the following sentences:
   a) Male sex cells are called i) __________ and female sex cells are called ii) __________.
   b) __________ is the joining together of the nuclei of male and female sex cells.
   c) A chromosome carries a line of __________.
   d) Alternative forms of a gene are called __________.
   e) One __________ allele will show in the offspring.
2. What two factors could cause two plants of the same species to look different?
3. Identical twins have the same genetic information. Why might they look different when they are older?

Producing plants and animals with the characteristics we prefer

Taking cuttings

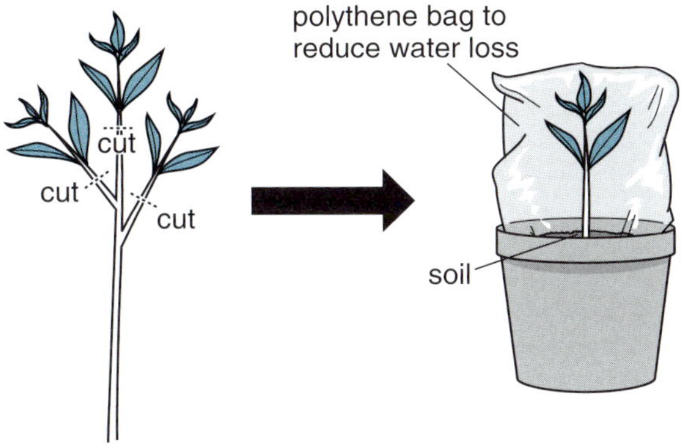

Take a young shoot from a plant and insert into a pot of soil.

The cutting will grow roots and form a new plant genetically identical to the parent plant.

Figure 4.4 Growing from a cutting

Modern cloning techniques

Cloning plants using tissue culture

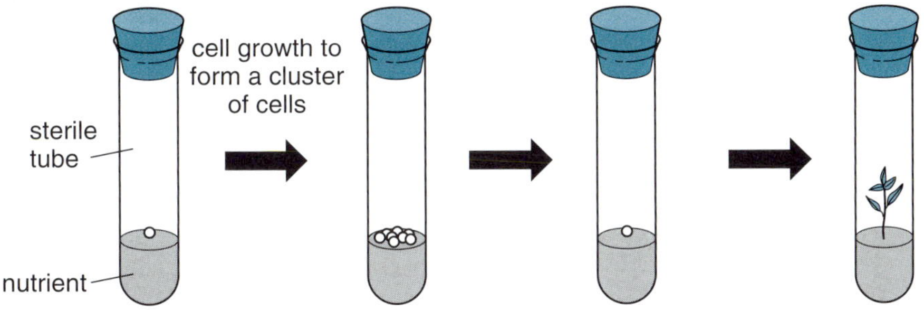

Each cell grows roots and shoots and forms a miniature plant. These can be grown on to full size plants. Each is a **clone** of the original donor plant, i.e. is genetically identical.

A cell is removed from the stem and placed on nutrient in a sterile tube.

Cells are separated and placed on new sterile nutrient with growth hormones. They are placed in warm, light conditions.

Figure 4.5 Tissue culture

Cloning animals by nuclear transfer

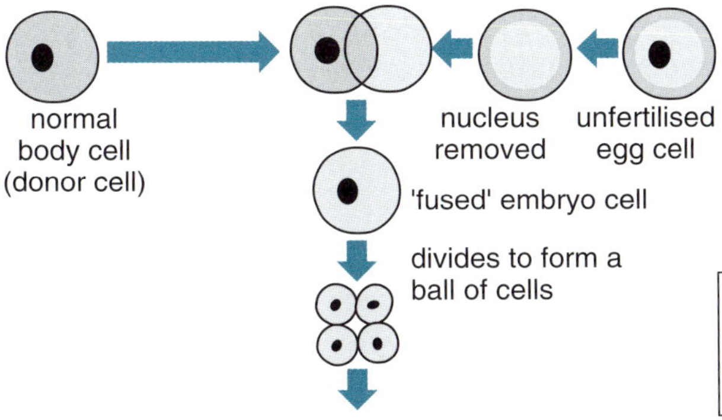

Donor cell is transferred to empty cell and joined by an electric shock.

The cells are implanted into an adult uterus to develop.

Figure 4.6 Nuclear transfer

Selective breeding

Selective breeding is a way of creating plants and animals which display a particular characteristic. For example, in cereal breeding we might choose 'large numbers of seeds' from one strain and 'drought resistance' from another strain. They could be cross pollinated in the hope of producing a strain that produced a large number of seeds even during a very dry summer.

Selective breeding creates problems by reducing the variety of alleles in a population. As a result:

- there are fewer alleles in the gene pool for further selective breeding
- the new strain produced may not survive if the environmental conditions change, i.e. wetter summers.

> **REMEMBER –** For selective breeding to be possible there must be genetic variation in the population to provide the alternative alleles so the breeder can select the useful characteristics. We need 'gene banks' which are collections of living plants and seeds that contain all the genetic variations.

Genetic engineering

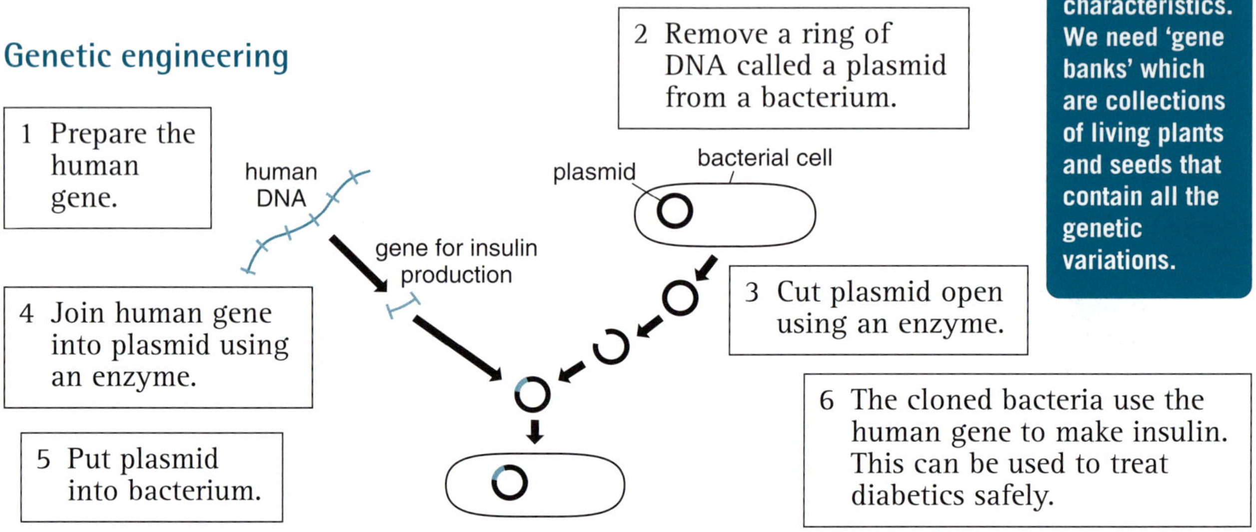

Figure 4.7 Using genetic engineering to produce human insulin

Using the human insulin gene in the bacteria means that the insulin produced is a copy of human insulin and so is both safe to use and cheap to produce.

CHECK YOUR UNDERSTANDING

1. Describe the flowers you would expect to grow if cuttings were taken from a red-flowered plant.
2. If a cell is taken from a plant and many are cloned using tissue culture, would you expect to see variation?
3. What problems arise from selective breeding?
4. What is the advantage of selective breeding?
5. Insulin produced by genetic engineering saves many lives. What is the advantage of producing it this way?

The theory of evolution

Fossils

If environmental conditions change, some plants or animals may not be able to survive in the new conditions and may become extinct (see page 22). Traces of some animals which died millions of years ago can be found as fossils but the majority would have been eaten or decomposed, leaving no trace.

a) Dinosaur's bones and teeth found in layers of mud deposits or leaves in coal.

b) Peat-bog man – acid water, no oxygen so decomposing bacteria could not survive.

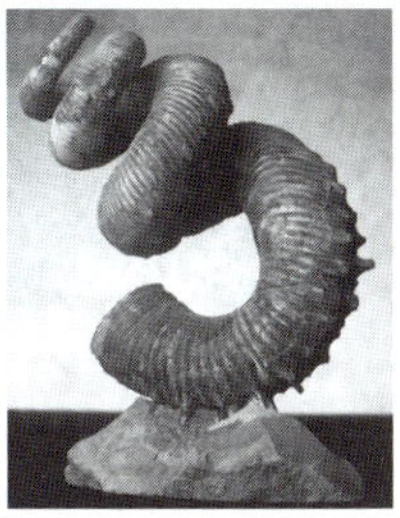

c) Whole organisms buried in layers of silt/chalk – mineral salts seep into them and form stone.

d) Dinosaur's footprints in mud.

Figure 4.8 Different ways in which fossils are formed

The theory of evolution states that all species of living things which existed in the past and those still present today evolved from simple water life forms which first developed more than 3 billion years ago.

What evidence do fossils provide?

Fossils show the order in which organisms evolved. Going down a cliff face, the rocks contain different fossils in different layers.

- fossils of mammals and flowering plants are found closest to the surface
- birds
- dinosaurs in layers 60 million years old
- reptiles
- fish
- insects are found in layers 350 million years old
- marshy land plants with no vascular support
- the oldest rocks are in the deepest layers and these have only fossils of simple marine (sea) life forms.

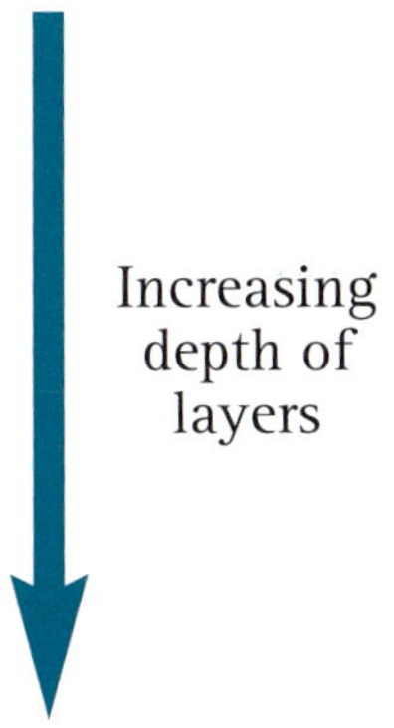

There are no living forms of some of the fossils and so these are said to be extinct, for example dinosaurs and woolly mammoths. They may have become extinct due to changes in the climate, or because of more successful new competitors, new predators or new diseases.

How do new species of plants and animals develop?

If changes occur to the DNA, this alters genes which in turn can change characteristics. If these alterations make the animal better adapted to the environment, the genetic differences may result in evolution of a new species after many generations and a long period of time.

What drives evolution?

Evolution is driven by **natural selection** which can work in the following way:

- Organisms produce more young than the environment can support so there is competition for food etc. (see page 23). Many organisms do not survive.
- There is **variation** in the population because individuals have different alleles. Some of these genetic differences result from **mutations**.
- Individuals with those alleles that give them the best adaptations to the environment survive, reproduce and pass the beneficial alleles on to the next generation.
- Well-adapted individuals rear more young.

Mutations

Mutations result from changes to the DNA resulting in new alleles. Some environmental factors can increase the mutation rate. These include exposure to ionising radiation, UV sunlight, powerful X-rays and certain chemicals.

Mutation and natural selection

Bacteria can be used to demonstrate 'speeded up evolution'.

- Bacteria can reproduce rapidly in ideal conditions, giving many generations in a single day.
- Mutations can occur as the DNA replicates for cell division.
- Some mutations change the bacterial chemistry so the bacteria are able to break down certain antibiotics i.e. the bacteria destroy the antibiotic rather than the antibiotic destroying them.
- Overuse of one antibiotic increases the likelihood that a clone of bacteria resistant to that antibiotic will arise by natural selection over many generations.

CHECK YOUR UNDERSTANDING

1. Will the oldest fossils be found at the top or the bottom of a cliff?
2. Sunburn may result in skin cancer. What has changed inside the cells to cause this?
3. A strain of bacteria has become resistant to the antibiotic penicillin. What does this mean?
4. Put the following fossils in order of age with the oldest first: mammals, fish, insects, birds, dinosaurs, reptiles, simple life forms.
5. Suggest why woolly mammoths became extinct.

Chromosomes and genes

- A chromosome is a long molecule of DNA.
- The DNA molecule is divided into many genes.
- A gene may control a characteristic.
- Some genes have alternative forms called alleles. For example, B and b are the alleles for black and brown fur in mice.
- An allele which controls the development of a characteristic when it is present on only one chromosome of a pair is a **dominant allele**. This is shown by a capital letter, e.g. B.
- If two matching alleles have to be present for a characteristic to show, one on each chromosome of the pair, the allele is **recessive**. A recessive allele is shown by a small letter, e.g. b.

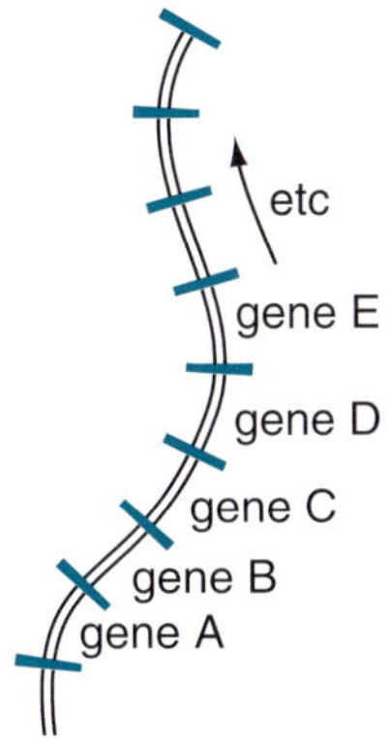

Figure 4.9 A chromosome with genes shown

Gender determination

Human body cells have **23 pairs of chromosomes**. One pair is the sex chromosomes which control your gender. Females have two X chromosomes; males have a non-matching pair of one X and one Y chromosome.

	Father	Mother	
	XY	XX	
Sperm gametes will contain →	X or Y	X	← All mother's eggs contain
Offspring (when sperm joins egg) which gives →	XX	or XY	
	50% females and 50% males		

Genetic diseases

Huntington's disease

Huntington's disease is a disorder of the nervous system caused by a dominant allele, H. The normal allele, h, is recessive.

	Parent suffering from Huntington's	Parent with normal alleles
	Hh	hh
Gametes (sperm or egg)	H or h	h
Children	50% Hh (sufferers)	50% hh (normal)

Only the parent with the dominant allele could pass the disease on to the children.

Cystic fibrosis

Cystic fibrosis is a disorder of the cell membranes and causes very sticky mucus all the time, resulting in breathing problems. To have this condition, the child must have two recessive alleles, so one must come in the sperm and the other in the egg. This means that father and mother are both Cc – they carry the recessive allele, c, but it is hidden by the dominant normal allele, C.

	Father (normal)	Mother (normal)
	Cc	Cc
Gametes	C or c	C or c

A punnet square is the easy way to work out all the combinations!

	C	c
C	CC	Cc
c	Cc	cc

1 in 4 of the children will receive two recessive alleles and will suffer from cystic fibrosis.

If you have to use a letter where capital and small can look the same, do make sure that there is a difference when you write them for dominant and recessive alleles.

Sickle cell anaemia

Sickle cell anaemia is a disorder of the red blood cells which reduces the oxygen-carrying capacity of the blood. The red blood cells become the shape of a new moon or sickle.

- A person who has two dominant alleles has normal oxygen transport but is likely to catch malaria.
- A person who has two recessive alleles has very poor oxygen transport and their muscles feel painfully 'cramped' when they use them.
- A person who has one dominant and one recessive allele can transport enough oxygen without pain and is less likely to catch malaria. But having both dominant and recessive alleles is an advantage only if you live near a tropical malarial swamp.

CHECK YOUR UNDERSTANDING

1. Does a father's sperm or a mother's egg determine the sex of the baby?
2. Does a father pass on his Y chromosome to his son or to his daughter?
3. What colour fur would you expect for the following mice: a) BB, b) Bb and c) bb?
4. If a father had the alleles Cc and a mother had cc, would you expect any of their children to suffer from cystic fibrosis?
5. Which pair of alleles would give you an advantage if you lived in a region where malaria was common?

Controlling fertility

Hormones in the menstrual cycle

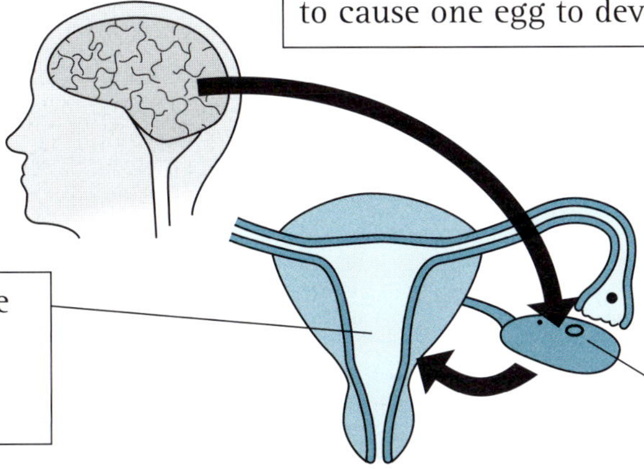

Figure 4.10 The hormones involved in the menstrual cycle

Controlling fertility using artificial hormones

If the brain produces too little stimulating hormone, an artificial form can be given to stimulate more eggs to mature. This increases the chance of an egg being fertilised but there is more risk of having more than one baby.

To prevent a pregnancy, 'oral contraception' (the birth pill) can be used. This prevents the development and release of eggs from the ovary. However, the pill gives no protection against sexually transmitted infections (STIs).

CHECK YOUR UNDERSTANDING

1. Which organ controls the monthly female cycle?
2. The brain produces a hormone at the start of the cycle – what does the hormone do?
3. What causes ovulation?
4. Which hormone prepares the uterus lining for a fertilised egg?
5. What is the risk of using a hormone to increase female fertility?

Topic Test

1 Copy and complete the following sentences.

a) The sperm (23 chromosomes) combines with the ovum (i) ____________ chromosomes) to form the cell which develops into the embryo. Each cell in the embryo contains ii) ____________ chromosomes.

b) As the embryo grows, more cells are produced by ____________ division.

c) A strawberry plant reproduces asexually by producing runners which grow a new plant at the end. The new plants will be genetically i) ____________ to the parent and to each other. They are called a ii) ____________ .

(5 marks)

2 Complete the following statements by matching up the correct two parts.

a) In the female the sex chromosomes	i) cannot be passed on to the next generation.
b) A mutation is a change in the DNA	ii) if two are present.
c) A mutation in a body cell	iii) are a matched pair.
d) Recessive alleles are only shown in an organism	iv) found on a chromosome.
e) A gene is a section of DNA	v) which can be caused by X-rays.

(5 marks)

3 A rat had a mutation and was able to eat rat poison without being killed. It mated with a normal rat and of the babies, half were like the mutant parent and were not killed by eating the poison.

a) Did this mutation result in a dominant or recessive allele?

b) Draw a genetic diagram to show the cross between the mutant rat and a normal rat.

c) Two of the surviving babies mated and had a litter of eight. Draw a genetic cross to explain how many of the eight would not be killed by the rat poison.

d) A pupil said that this was an example of 'natural selection'. What did he mean? (9 marks)

4 a) Explain the two ways in which hormones can be used to alter a woman's fertility.

b) What problems could arise from using hormones to control fertility? (3 marks)

5 a) Copy and complete the following sentences.

To produce human insulin using genetically engineered bacteria, first the i) ____________ has to be obtained from the human pancreas cell. A small circular piece of DNA is removed from the ii) ____________ cell and cut open using an iii) ____________ . The human gene is joined into the bacterial DNA and inserted back into the bacterial cell which then makes insulin.

b) Give two advantages of making human insulin for the patient. (5 marks)

6 Two parents who are normal have a child who suffers from cystic fibrosis.

a) What is the genotype of i) the mother ____________, ii) the father ____________ ?

b) If this couple were to have another child, what is the probability that the second child would not suffer from cystic fibrosis? (3 marks)

Total: 30 marks

Biology examination questions and model answers

1 The diagram shows one of the experiments performed by a scientist called Mendel. He bred sweet pea plants.

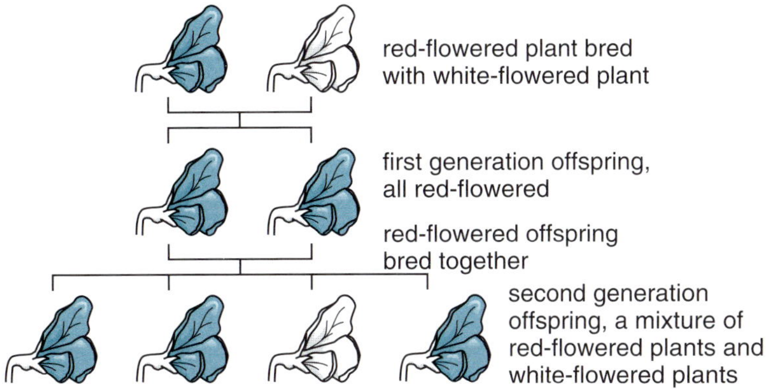

In the sentences below, cross out the **two** lines which are wrong in each box. *(5 marks)*

Mendel proposed that flower colour was controlled by inherited factors.

The first generation plants show that the red factor is

dominant
~~environmental~~
~~recessive~~

The second generation plants show that the white factor is

~~dominant~~
~~environmental~~
recessive

We now call inherited factors

~~chromosomes~~
~~gamates~~
genes

These factors are passed from generation to generation in

gametes
~~glands~~
~~organs~~

The red-flowered sweet pea plants did not all grow to the same height.

This was due to

~~dominant~~
environmental
~~recessive~~

factors.

> Each sentence is worth one mark.
> Make sure you clearly cross out **two** lines in each box.

2 a) The question describes the process of photosynthesis.

carbon dioxide + water + light energy ⟶ glucose + oxygen

(i) Write in the names of the **two** missing substances in the equation above. *(2 marks)*

There are two missing substances each worth one mark. So don't write more than one in each gap.

(ii) Name the green substance which absorbs the light energy. *(1 mark)*

chlorophyll

There is only one correct answer. Do not write chloroplast, it is not a substance.

b) (i) In bright sunlight, the concentration of carbon dioxide in the air can limit the rate of photosynthesis. Explain what this means. *(2 marks)*

There is enough light to make photosynthesis go faster, but there is not enough carbon dioxide in the air to speed photosynthesis up. Both are needed.

The need for both light and carbon dioxide has to be given.

(ii) Give one environmental factor, other than light intensity and carbon dioxide concentration, which can limit the rate of photosynthesis. *(1 mark)*

temperature

Rainfall would be an acceptable alternative answer.

METALS

Metals in the Periodic Table

The chemical elements can be arranged in order of their **relative atomic mass** (A_r).

KEY: gases | metals | non-metals (solid or liquid)

Symbol	H	He	Li	Be	B	C	N	O	F	Ne	Na	Mg	Al	Si	P	S	Cl	Ar	K	Ca
A_r	1	4	7	9	11	12	14	16	19	20	23	24	27	28	31	32	35.5	40	39	40

(Potassium (K) and argon (Ar) have been reversed to improve the 'fit'.)
This can be rearranged to form the **Periodic Table**.

Period \ Group	1	2											3	4	5	6	7	0
								H										He
2	Li	Be											B	C	N	O	F	Ne
3	Na	Mg	transition elements (metals)										Al	Si	P	S	Cl	Ar
4	K	Ca	Sc	Ti	V	Cr	Mn	Fe	Co	Ni	Cu	Zn	Ga	Ge	As	Se	Br	Kr
5	Rb	Sr	Y	Zr	Nb	Mo	Tc	Ru	Rh	Pd	Ag	Cd	In	Sn	Sb	Te	I	Xe
6	Cs	Ba	La	Hf	Ta	W	Re	Os	Ir	Pt	Au	Hg	Tl	Pb	Bi	Po	At	Rn
7	Fr	Ra	Ac															

Figure 5.1 The Periodic Table

The table below compares some of the properties of two groups of metals – the alkali metals and the transition metals.

	Group 1 Metals (Alkali Metals)	Transition Metals
Properties common to all metals	Good conductors of heat Good conductors of electricity Easy to bend or hammer into shape	
Physical properties	Low melting point (Na = 98°C) Low density (Li, Na and K float on water) Soft, easily cut with a knife	High melting point (except mercury which is a liquid at room temperature) (Fe = 1539°C) High density Hard, strong, tough

Chemical properties	Very reactive Form ionic compounds with non-metals (e.g. Li_2O, NaCl) Compounds are white solids, they dissolve in water to give colourless solutions React with water (metal + water → hydrogen + metal hydroxide) Metal hydroxides are soluble in water to form alkaline solutions	Not very reactive Do not corrode easily in air or water, so they are used as structural materials (e.g. steel) and for electrical wiring (e.g. copper) Form coloured compounds, so they are used as glazes to colour pottery Used as catalysts (e.g iron in the manufacture of ammonia and platinum to make nitric acid from ammonia

CHECK YOUR UNDERSTANDING

❶ What does the symbol A_r represent?

❷ With which element could the symbol A_r be confused?

❸ Approximately what percentage of the elements are metals?

❹ Give three ways in which the physical properties of the Group 1 metals differ from those of the transition metals.

❺ Give three ways in which the chemical properties of the Group 1 metals differ from those of the transition metals.

Extraction of metals from their ores

The reactivity series

Metals can react:

1 With air to give metal oxides.
2 With water (cold, hot or as steam) to produce metal oxides (or hydroxides) and hydrogen gas.
3 With acids to produce metal salts and hydrogen gas.

From observing such reactions, we can draw up a reactivity series of metals as shown in Figure 5.2.

Once we have this information, we can use it to make predictions about the outcome of reactions, as a more reactive metal (and the non-metals carbon and hydrogen) can displace a less reactive metal from its compounds.

So: magnesium + copper oxide → magnesium oxide + copper

$Mg + CuO \rightarrow MgO + Cu$

sodium + water → sodium hydroxide + hydrogen

$2Na + 2H_2O \rightarrow 2NaOH + H_2$

iron + copper sulphate solution → iron sulphate + copper

$Fe(s) + CuSO_4(aq) \rightarrow FeSO_4(aq) + Cu(s)$

But: gold + water → NO REACTION

silver + copper oxide → NO REACTION

potassium
sodium
calcium
magnesium
aluminium
carbon*
zinc
iron
tin
lead
hydrogen*
copper
silver
gold
platinum

↑ Increasing reactivity with air, water or acids

Figure 5.2 The reactivity series of metals.
*Carbon and hydrogen are not metals. They are included for comparison

Metals are found in rocks in the Earth's crust. Some unreactive metals like gold occur as the metal itself. Most metals are found as metal compounds – usually metal oxides. Most metal ores are oxides (or substances that can easily be changed into metal oxides). To extract a metal from its oxide, the oxygen must be removed. This is an example of **reduction**.

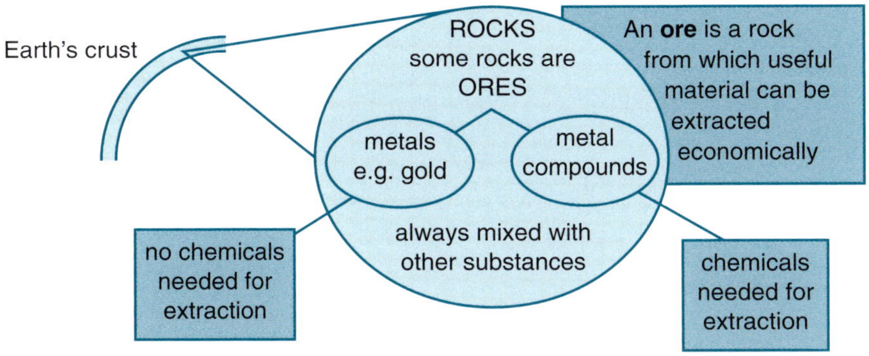

Figure 5.3 Extracting metals from the Earth's crust

How a metal is extracted from its ore depends on the position of the metal in the reactivity series. Many metals can be extracted from their compounds by using a more reactive 'metal'. To be cost effective the substance used must be cheap and available. So **carbon** is used to extract metals (like iron) that are lower down the reactivity series. The process is called **smelting**.

Position of metal in reactivity series	Method of extraction
Metals at the bottom e.g. gold	Occur naturally as the metal itself
Metals below carbon e.g. iron	Heating with carbon
Metals above carbon e.g. aluminium	Electrolysis

CHECK YOUR UNDERSTANDING

1. Copy and complete these word equations. (*Hint*: Not all will react.)
   a) zinc + lead oxide
   b) tin + aluminium chloride solution
   c) hydrogen + copper oxide
2. What is the difference between a rock and an ore?
3. Why is carbon used to extract metals like iron rather than using a more reactive metal like magnesium?

Extracting iron

Iron is extracted from its ore in the **blast furnace** by heating it with carbon (coke).

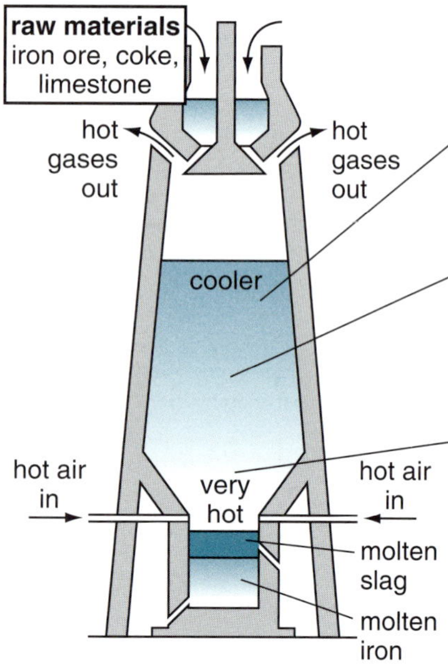

iron ore + carbon monoxide → iron + carbon dioxide

$$Fe_2O_3 + 3CO \rightarrow 2Fe + 3CO_2$$

(CO oxidised to CO_2; Fe_2O_3 reduced to Fe)

coke + carbon dioxide → carbon monoxide

$$C + CO_2 \rightarrow 2CO$$

(C oxidised; CO_2 reduced)

coke burns in hot air

$C + O_2 \rightarrow CO_2$ (+ energy, which keeps the furnace hot)

Figure 5.4 The reactions in a blast furnace

In the blast furnace coke burns in the hot air blast, producing carbon dioxide. The reaction gives out a lot of heat. The carbon dioxide then reacts with more carbon to produce **carbon monoxide**. This carbon monoxide reduces iron ore (iron oxide or **haematite**) to iron. In this process the carbon monoxide is oxidised to carbon dioxide.

When a substance joins with oxygen, we call it an **oxidation** reaction. All ores contain impurities. In iron ore the main impurity is sand (silicon dioxide). The impurity must be removed or it would weaken the iron produced. The sand reacts with the limestone added as part of the original charge, producing **slag**.

$$CaCO_3 + SiO_2 \rightarrow CaSiO_3 + CO_2$$

limestone + sand → slag + carbon dioxide

The slag melts and settles on top of the molten iron at the bottom of the blast furnace.

CHECK YOUR UNDERSTANDING

1. What happens to the oxygen content of iron ore when it is reduced?
2. What happens to the oxygen content of carbon monoxide when it is oxidised?
3. a) Why does the impurity have to be removed from the iron ore?
   b) How is the impurity removed?

Electrolysis

Aluminium is too reactive to be extracted by smelting. A more powerful method is used. The process is called electrolysis.

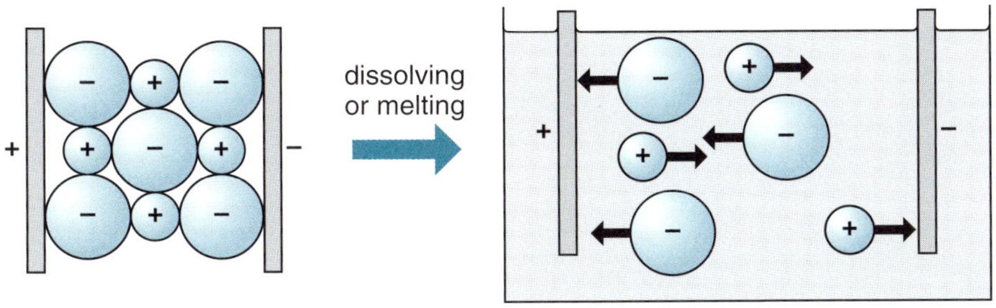

In an ionic solid, the ions are held in place by each other.

Because they cannot move, the solid does not conduct electricity.

When the ionic substance is molten or dissolved in water, the ions are free to move.

They are attracted to the electrode with the opposite charge. Positive ions move towards the cathode (negative electrode) and negative ions move towards the anode (positive electrode).

Because the charges move, the ionic substance now conducts electricity.

Figure 5.5 Electrolysis

- When negative ions reach the anode they lose electrons and become atoms.
- When positive ions reach the cathode they gain electrons and become atoms.
- So electrolysis causes ionic substances to break down (decompose). The process can produce gases (mainly at the anode) or metals (at the cathode).
- If the ionic substance is aluminium oxide, the reactions which occur are:

  At the anode:
  Negatively charged oxide ions lose electrons to become oxygen gas.

  At the cathode:
  Positively charged aluminium ions gain electrons to become aluminium metal.

Extracting aluminium

Aluminium is extracted by electrolysis using a cell like that in Figure 5.6.

The purified aluminium ore (bauxite) is dissolved in molten cryolite (another aluminium ore). This is because the melting point of bauxite alone is very high.

Aluminium forms at the cathode. Oxygen forms on the carbon anode. The oxygen reacts with the carbon to form carbon dioxide. This causes the anodes to burn away so they have to be replaced frequently.

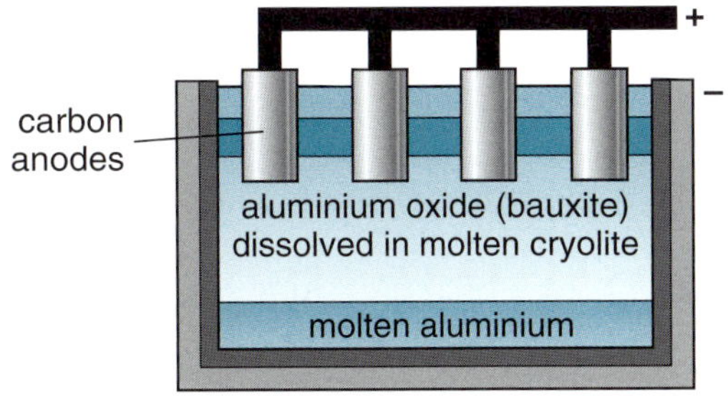

Figure 5.6 Extracting aluminium by electrolysis

Purifying copper

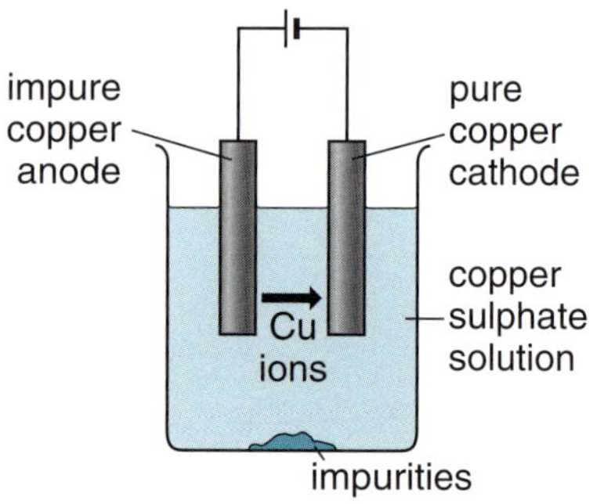

Figure 5.7 Purifying copper

- The positive electrode (anode) is impure copper.
- The negative electrode (cathode) is pure copper.
- The solution contains copper ions (positive).
- The copper ions in the solution go to the cathode and become copper atoms.
- The copper atoms join to the pure copper cathode.
- The cathode gets thicker.
- The copper in the impure anode forms copper ions which replace those removed from the solution.
- The impure anode gets thinner.
- Impurities from the anode fall to the bottom.

CHECK YOUR UNDERSTANDING

1. Why don't solid ionic compounds conduct electricity?
2. Why do ionic compounds conduct electricity when molten or dissolved in water?
3. What are the name and formula of the main ore of aluminium?
4. a) At which electrode in the cell does aluminium form?
   b) What forms at the other electrode of the cell?
   c) Why do the carbon anodes have to be replaced quite often?

Preventing corrosion

Iron and steel

Iron and steel corrode (rust) more rapidly than most transition metals. Adding less reactive metals like chromium or nickel can make non-rusting **alloys**. These alloys are known as **stainless steel**.

Iron and steel can also be protected by **sacrificial protection**.

In this process magnesium or zinc is used. Blocks of either metal are fixed to the iron object (e.g. a ship). Because iron is less reactive, the magnesium or zinc corrodes first.

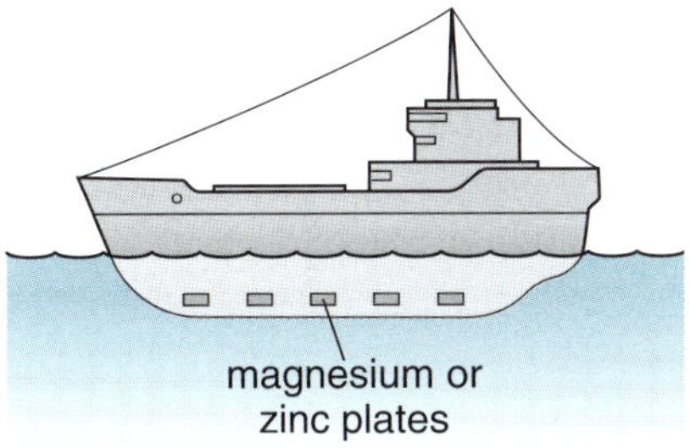

Figure 5.8 Sacrificial protection

Aluminium

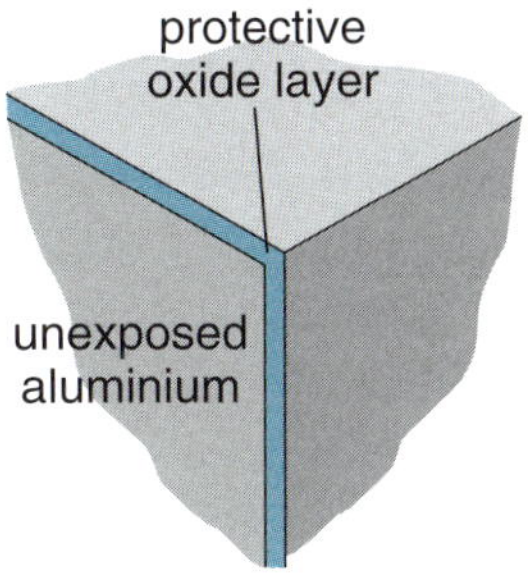

Figure 5.9 A section through a block of aluminium

Aluminium is quite high up in the reactivity series. But aluminium does not seem to be very reactive. (You can boil water in aluminium saucepans for years without them corroding.) This is because aluminium reacts with oxygen in the air to form aluminium oxide. The aluminium oxide forms a thin, protective layer on the surface. It seals the unexposed aluminium so that air and water cannot reach it.

> Aluminium can be made harder, stiffer and stronger by adding other elements. Magnesium is an element that can be used. These aluminium alloys are useful as structural metals.

CHECK YOUR UNDERSTANDING

❶ What word is used to describe the corroding of iron?

❷ a) Which elements are used to form a non-corroding alloy with iron?
b) What is this alloy called?

❸ a) Which two metals are used to protect iron and steel by being fixed to the surface?
b) What property must these metals have to make this process work?
c) What is the process called?

❹ Why is aluminium less reactive than its position in the reactivity series suggests?

Making metal compounds

Acids and alkalis

All **acids** contain **hydrogen ions** (H^+). It is the hydrogen ion that makes the solution acidic. All **alkalis** contain **hydroxide ions** (OH^-). It is the hydroxide ion that makes the solution alkaline.

When an acid reacts with an alkali the process is called **neutralisation**. The products are a **salt** and water.

acid + alkali → salt + water

LEARN this word equation.

Salts

Salts have a metal 'bit' (from the alkali) and an acid 'bit'. The name of the acid is changed when it forms a salt.

The salt produced depends on the acid used and the metal in the alkali.

sodium hydroxide + hydro**chloric** acid → sodium **chloride** + water

$NaOH + HCl \rightarrow NaCl + H_2O$

Hydrochloric acid produces chloride salts.

potassium hydroxide + **sulph**uric acid → potassium **sulphate** + water

$2KOH + H_2SO_4 \rightarrow K_2SO_4 + 2H_2O$

Sulphuric acid produces sulphate salts.

calcium hydroxide + **nitr**ic acid → calcium **nitrate** + water

$Ca(OH)_2 + 2HNO_3 \rightarrow Ca(NO_3)_2 + 2H_2O$

Nitric acid produces nitrate salts.

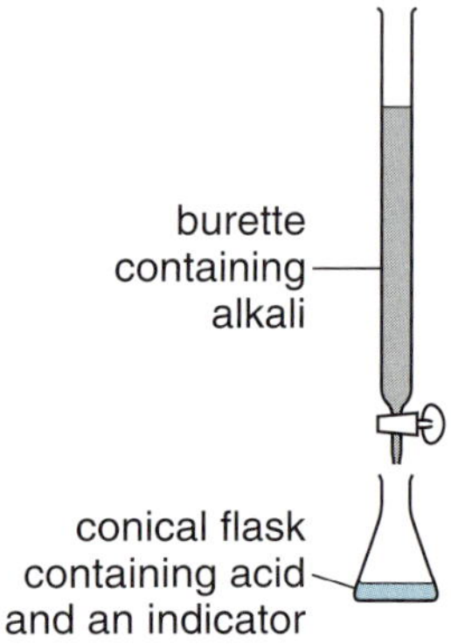

Figure 5.10 Neutralising an acid with an alkali

Ammonia (NH_3) also forms alkaline solutions. When neutralised by an acid it forms **ammonium** salts.

ammonia + hydrochloric acid → ammonium chloride

The oxides and hydroxides of transition metals are not soluble in water. They are called **bases**. An alkali is also a base, but one that is soluble in water.

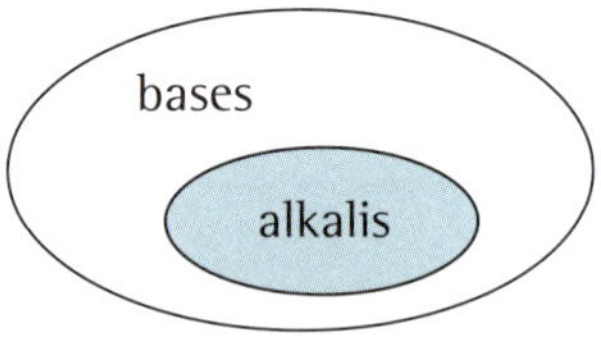

Figure 5.11 An alkali is a soluble base

Salts of transition metals can be made by adding the transition metal oxide (or hydroxide) to the acid. The base is added in small quantities until no more will 'dissolve' (this means the reaction is complete). All the acid has now been neutralised. The excess base is filtered off.

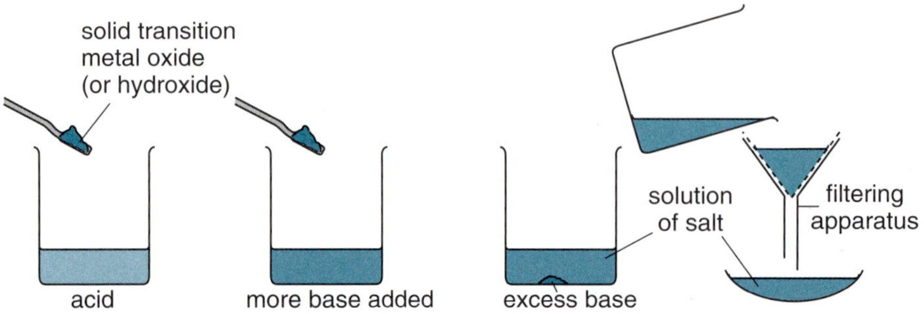

Figure 5.12 Making a transition metal salt

CHECK YOUR UNDERSTANDING

❶ What is the pH of water?

❷ What substances are used to see whether a solution is acidic or alkaline?

❸ When an acid reacts with an alkali, a salt is formed. What else is always formed?

❹ What salt is formed in the following reactions?
a) Hydrochloric acid and potassium hydroxide.
b) Lithium hydroxide and nitric acid.
c) Ammonia solution and sulphuric acid.

❺ What is the difference between a base and an alkali?

Topic Test

1 This question is about elements. Match each element in the box with the statements numbered (i) to (iv).

carbon	chlorine	iron	mercury

(i) It is a gas.
(ii) It is a liquid.
(iii) It is a solid metal.
(iv) It is a solid non-metal. (4 marks)

2 Which two of the following symbols represent transition metals?

Mg Mn Pb Rb Ru (2 marks)

3 Which two elements in the box would react with iron oxide?

carbon	copper	lead	magnesium	tin

(2 marks)

4 In the blast furnace the following substances are added.

coke, hot air, iron ore, limestone

Use the substances in the list to answer the questions below.

a) Which substance burns to keep the furnace hot?
b) Which substance is **not** added at the top of the furnace?
c) Which substance is used to convert impurities into slag?
d) Which substance reacts with carbon monoxide in the blast furnace? (4 marks)

5 This question is about electrolysis. Select the correct answer from each list.

a) Ionic compounds conduct electricity when they are molten. When else can they conduct electricity?
  (i) If they are soluble solids.
  (ii) If they are insoluble solids.
  (iii) When they are in solution.
  (iv) Either as a solid or in solution.

b) In the electrolysis of a molten ionic substance, which ions travel to the positive electrode (anode)?
  (i) Positive ions.
  (ii) Negative ions.
  (iii) Both positive and negative ions.
  (iv) None of the ions.

c) At the positive electrode ions
  (i) deposit as metals.
  (ii) gain electrons.
  (iii) join together.
  (iv) lose electrons.

d) Electrolysis of a molten ionic substance causes it to
  (i) be neutralised.
  (ii) break down.
  (iii) combine.
  (iv) turn into a solution. (4 marks)

6 For each of the following state if the process uses electrolysis or heating with carbon:

a) extracting aluminium

b) extracting iron

c) purifying copper. (3 marks)

7 Name two ways in which steel can be protected from corrosion. (2 marks)

8 Aluminium is high in the reactivity series yet it can be used to make saucepans. Explain this. (3 marks)

9 Name a substance that would have pH of 4. (1 mark)

10 a) Copy and complete the table.

Acid	Base/Alkali	Salt
sulphuric acid	ii)	sodium sulphate
hydrochloric acid	potassium hydroxide	iii)
i)	copper oxide	copper nitrate

b) What type of reactions are these?

c) In each reaction a salt is formed. What else is formed? (5 marks)

Total: 30 marks

EARTH MATERIALS

Uses of limestone

Limestone is a very common substance, which makes it a relatively cheap material to use.

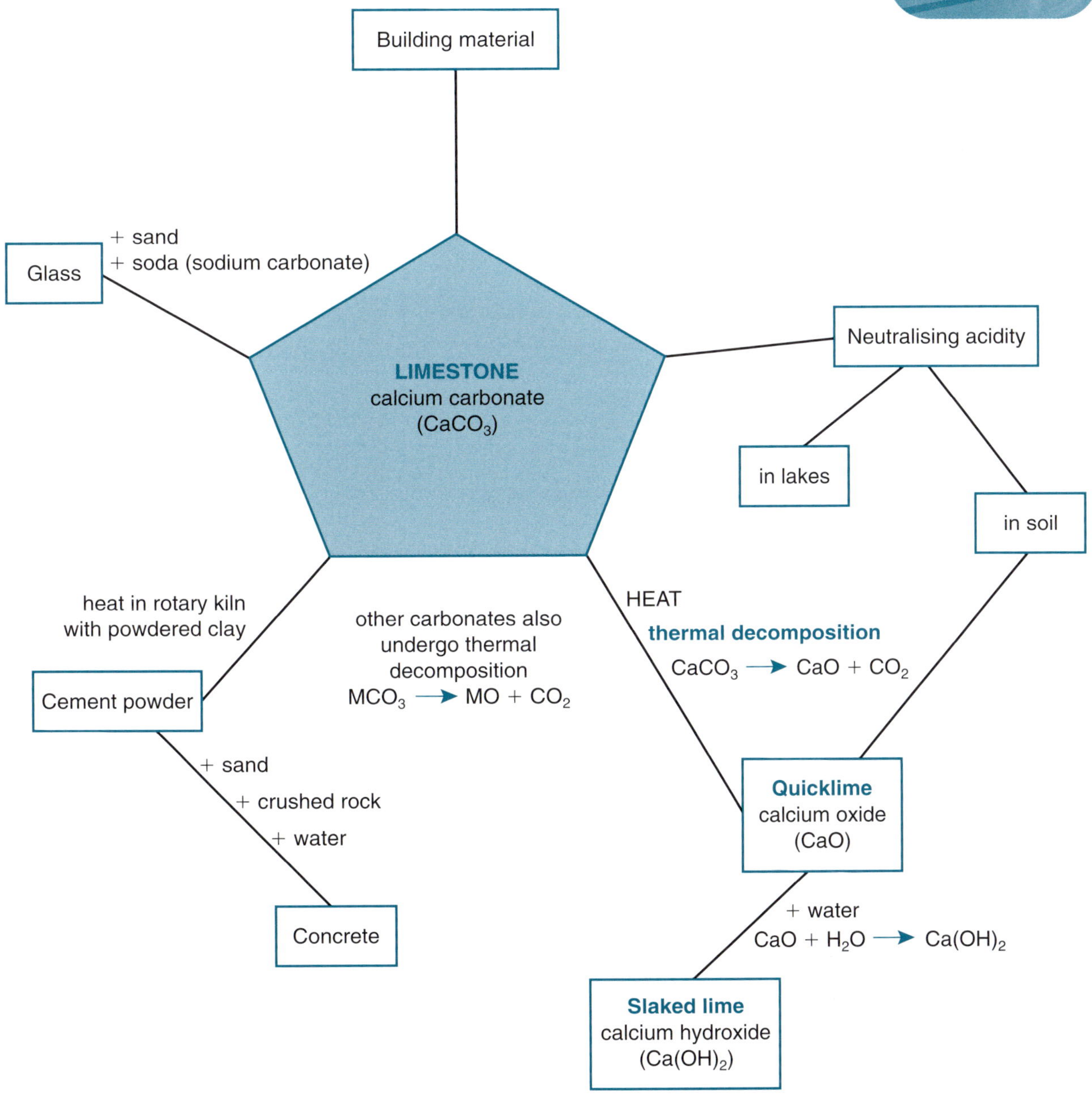

Figure 6.1 Uses of limestone

CHECK YOUR UNDERSTANDING

1. Write word equations for the following reactions of limestone and its products.
   a) The thermal decomposition of limestone.
   b) The reaction between quicklime and water.
   c) The reaction in which limestone neutralises sulphuric acid in lakes.
2. Why is quicklime not used to neutralise acidity in lakes?
3. Name one process not included in Figure 6.1 that uses limestone.

Useful products from crude oil

What is crude oil?

Crude oil is a mixture of different hydrocarbons.

A **mixture** contains two or more substances.
- The substances are not chemically combined.
- The chemical properties of each substance are unchanged.
- The different substances can easily be separated (e.g. by distillation).

- A **hydrocarbon** is a compound that contains only carbon and hydrogen.
- The different hydrocarbons in crude oil are not chemically combined and can be separated by fractional distillation.
- Within each fraction the molecules have about the same number of carbon atoms.

The fractional distillation of crude oil

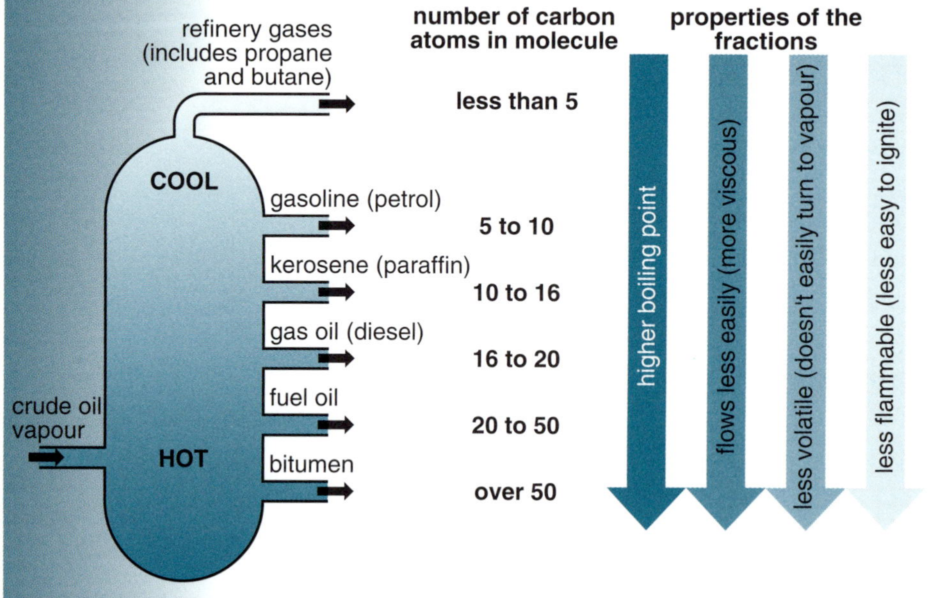

Figure 6.2 Separating the fractions of crude oil

The large hydrocarbon molecules are not very useful. They can be broken down into smaller, more useful molecules by **cracking**.

Cracking

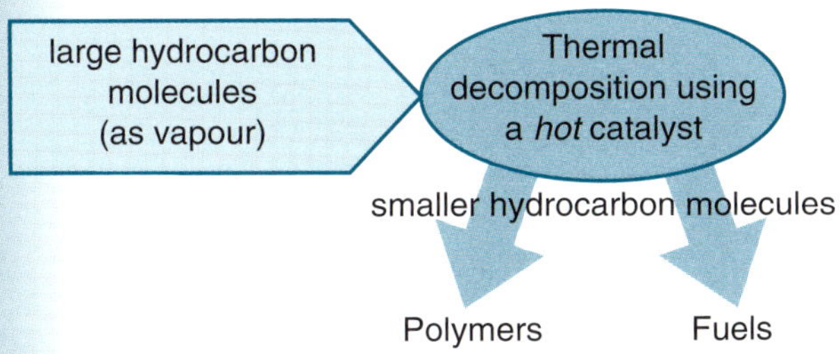

> **Cracking** is the process that breaks large hydrocarbon molecules into smaller ones.

Polymers include **plastics**.

Poly(ethene) is used for making plastic bags and bottles.

Poly(propene) is used for making crates (e.g. milk crates) and ropes.

Most plastics are not **biodegradable**. This means they are not broken down by micro-organisms. This causes problems if the plastics are buried in landfill sites.

Plastics can be recycled or incinerated (burned).

These fuels contain **carbon** and **hydrogen**.

Sulphur may also be present as an impurity.

When the fuel is burned, the following reactions take place:

carbon + oxygen → carbon dioxide
hydrogen + oxygen → water (vapour)
sulphur + oxygen → sulphur dioxide

Sulphur dioxide (SO_2) is a pollutant. It causes **acid rain**.

Carbon dioxide (CO_2) is a **'greenhouse' gas**.

CHECK YOUR UNDERSTANDING

1. What is meant by the word 'hydrocarbon'?
2. What is the process used to separate the different substances in crude oil?
3. List the main fractions into which crude oil is broken down.
4. Give five ways in which the fractions differ from each other.
5. Which fractions of crude oil are the least useful?
6. What is the name of the process that can convert the least useful fractions of oil into more useful products?
7. What are the two main uses of the products formed by cracking?
8. Explain why putting plastic waste into a landfill site is not environmentally friendly.
9. What methods other than landfill are there of disposing of plastics? What are the environmental effects of these methods of disposal?
10. When fuels are burned what are the two main products?
11. Fuels often contain sulphur as an impurity. What are the environmental effects of burning sulphur-containing fuels?

The creation of the Earth's atmosphere

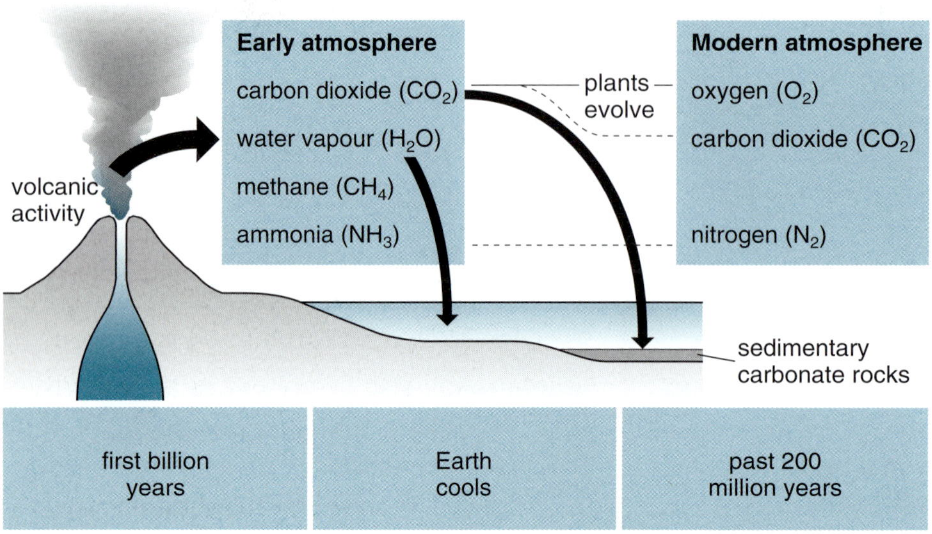

Figure 6.3 How the Earth's atmosphere was formed

Intense volcanic activity created the first atmosphere. It contained carbon dioxide, water vapour and small amounts of methane and ammonia. As the Earth cooled, the water vapour formed oceans. Some of the carbon dioxide dissolved in the oceans. Eventually it became 'locked up' in **sedimentary, carbonate rocks**.

At this stage the atmosphere was similar to that of Venus or Mars today – there was almost no oxygen.

Once green plants evolved, the process of **photosynthesis** began. Gradually the plants took up the carbon dioxide and oxygen was released. Many of the micro-organisms on Earth at that stage could not survive in the presence of oxygen. As the Earth became 'polluted' with oxygen, they died.

Some of the plants became fossilised and the carbon was again 'locked up' – this time in the form of **fossil fuels**. Methane and ammonia reacted with some of the oxygen to form carbon dioxide, water and nitrogen.

About 200 000 000 years ago the atmosphere reached its present composition. Since then it has not changed very much.

Composition of atmosphere now	
Gas	**%**
nitrogen	80
oxygen	20
other gases (water vapour, noble gases and carbon dioxide)	about 1

CHECK YOUR UNDERSTANDING

1. How was the Earth's original atmosphere formed?
2. When was the Earth's original atmosphere formed?
3. Name four gases present in the original atmosphere. For each gas, explain what happened to it as the atmosphere evolved.
4. What is the composition of the Earth's present atmosphere?

The structure of the Earth

The crust is very thin – between 10 km and 100 km thick.

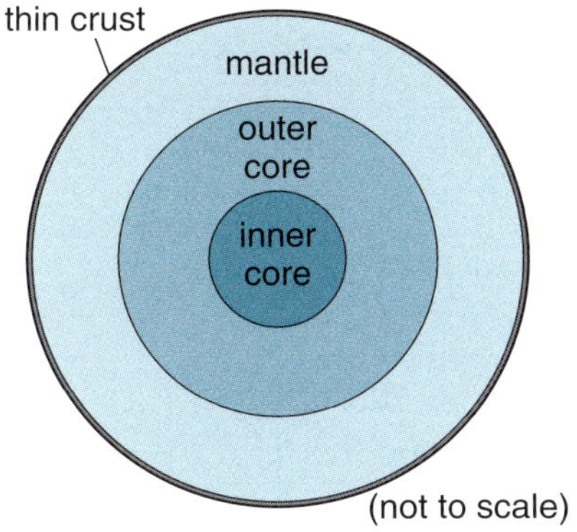

The radius of the Earth's core is just over half that of the Earth. It is made of nickel and iron. The inner core is solid and the outer core is liquid.

The mantle is made of rock. It has the properties of a solid except that it can flow very slowly (like very thick treacle).

Figure 6.4 A section through the Earth

The average density of the Earth (about 5.5 g/cm^3) is much higher than that of the crust (about 2.7 g/cm^3). This means that the Earth's core must have a much higher density (about 11 g/cm^3). This is because the Earth's core is made from a different material from the crust.

Rocks are continually being recycled. The process is very slow and takes many millions of years to complete.

CHECK YOUR UNDERSTANDING

1. Which part of the Earth has the highest density?
2. a) The Earth's internal structure has four main parts. What are these parts called?
   b) Which of these four parts has the largest volume?
3. What is unusual about the rock material that makes up the mantle?

Sedimentary rocks

Sedimentary rocks form from small fragments of rock (or the shells of dead sea creatures) that are deposited on the sea bed. The appearance of the sedimentary rock is evidence of how the fragments were deposited.

The rocks are usually found in layers. These layers formed because the deposits did not settle out in a regular way. In some cases the deposits have ripples that were caused by waves or currents in the sea.

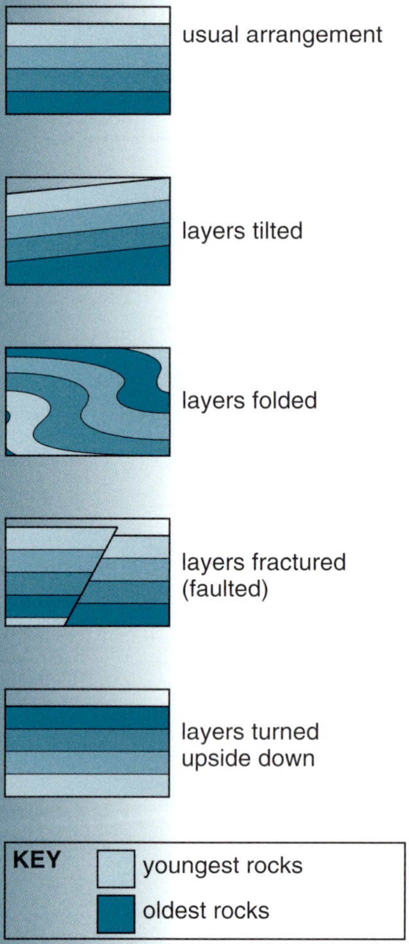

Usually the youngest rocks lie on top of older rocks.

The Earth's crust is not very stable. It is subject to very large forces. These forces can cause the crust to move and the layers to tilt. The movement of the crust is known as **tectonic activity**.

In very severe cases the layers might be folded on top of each other . . .

. . . or fractured and displaced. (The lines of the fracture are called **faults**.)

If the forces are large enough, the rocks might actually be turned upside down. When this happens, the oldest rocks are found on top.

Figure 6.5 The structure of sedimentary rocks

When the crust moves, mountain ranges are sometimes formed. The process is very slow – it can take millions of years. These new mountain ranges replace older ones which are being gradually worn down by **weathering** and **erosion**.

Metamorphic rocks

Metamorphic rocks are formed when sedimentary rocks are heated to high temperatures and/or put under great pressure. The Earth movements that can distort sedimentary rocks can also cause the temperatures and pressures needed.

CHECK YOUR UNDERSTANDING

1. Name four ways in which sedimentary rocks can be distorted.
2. What name is given to the movement of the Earth's crust?
3. How are metamorphic rocks formed?

Continental drift

The west coast of Africa and the east coast of South America are several thousand kilometres apart. But they seem to fit together like pieces in a huge jigsaw. The fossils and rock structures are also similar. These facts suggest that they were once part of the same land mass.

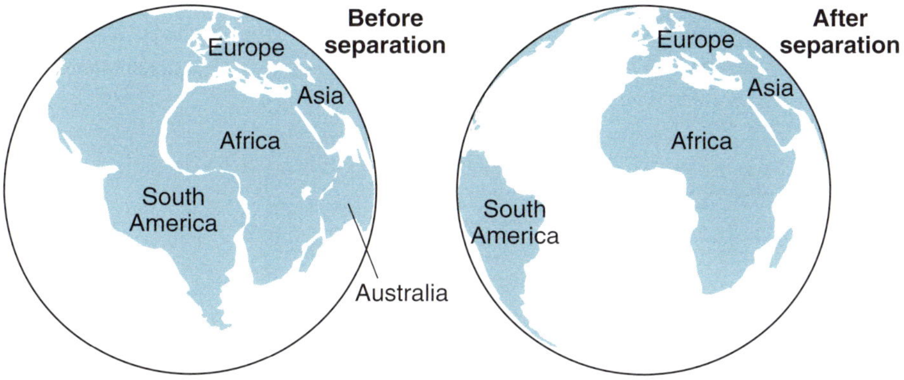

Figure 6.6 These maps show how South America and Africa may once have fitted together

In the early part of the twentieth century, Alfred Wegener put forward the idea of continental drift. He thought that the Earth's crust was moving. His ideas were not taken seriously. At that time scientists thought that the Earth was shrinking as it cooled down. They believed that mountains, earthquakes and volcanoes were caused by distortion of the Earth's crust as the core cooled and got smaller.

It was almost 50 years after Wegener proposed his theory before scientists discovered that the Earth's surface is moving. It moves very slowly – just a few centimetres each year. The Earth's lithosphere (the crust and the upper part of the mantle) is made up of a number of large pieces. These pieces are called tectonic plates.

Radioactive processes inside the Earth produce heat. This heat causes convection currents in the mantle. The tectonic plates move because of these convection currents. Earthquakes and volcanic activity occur at the boundaries between the plates.

The movements of the plates are very complex, so scientists are not able to predict exactly when earthquakes and volcanic eruptions will occur.

CHECK YOUR UNDERSTANDING

1. Name two pieces of evidence that suggested to scientists that South America and Africa were once joined.
2. a) What was Wegener's theory called?
   b) Why was it not accepted until 50 years after he had proposed it?
3. The Earth's lithosphere is made up of a number of plates.
   a) What is meant by the word lithosphere?
   b) What is the general term used to describe the plates?
   c) What causes the plates to move?

Topic Test

1 There are many ways in which limestone is used. Match the uses of limestone in the box with the correct method of treatment below.

building	cement	glass	quicklime

Limestone is:

a) heated strongly.
b) heated strongly with clay.
c) heated strongly with sand and soda.
d) used with no processing. (4 marks)

2 Which two of the following are NOT true for a mixture?

a) The substances present are chemically combined.
b) The chemical properties of each substance present are unchanged.
c) The different substances are usually easy to separate.
d) Air is an example of a mixture.
e) Water is an example of a mixture. (2 marks)

3 Crude oil can be separated into its components by fractional distillation. For which of the fractions in the table below are both statements true?

Fraction	Name of fraction	No. of carbon atoms in molecule	Property of the fraction
A	refinery gases	less than 5	have high melting points
B	petrol	5–10	is less flammable than diesel oil
C	bitumen	10–16	is almost black in colour
D	diesel oil	16–20	is more viscous than petrol
E	fuel oil	20–50	has a darker colour than diesel oil

(2 marks)

4 The following questions are about cracking hydrocarbon molecules. For each question pick the correct answer from the list.

a) Cracking is an example of:
   (i) combustion.
   (ii) fractional distillation.
   (iii) neutralisation.
   (iv) thermal decomposition.

b) Cracking can be used to make substances needed to make plastics. Which of the statements is true?
   (i) All plastics can be recycled.
   (ii) Plastics are biodegradable.
   (iii) Poly(ethene) is used to make milk crates.
   (iv) Poly(propene) is used to make ropes.

c) In the process of cracking:
   (i) large molecules are broken down into smaller ones.
   (ii) large molecules are combined to make polymers.
   (iii) small molecules are broken down into carbon atoms.
   (iv) small molecules are combined to make polymers.

d) Cracking can also produce fuels. Which statement is NOT true?
   (i) The fuels contain hydrogen and carbon.
   (ii) The products of combustion include carbon dioxide and water vapour.
   (iii) Sulphur may be present as an impurity.
   (iv) Sulphur burns to produce sulphur dioxide which can make rain alkaline. (4 marks)

5 Since the Earth formed, its atmosphere has changed.

a) Name two gases that were in the early atmosphere that are also present in the modern atmosphere.

b) Name two gases that are in the atmosphere now but were not in the early atmosphere. (4 marks)

6 a) Name the two metals that make up most of the Earth's core.

b) Which of the Earth's layers is the thinnest?

c) Which of the Earth's layers are mainly solid? (6 marks)

7 Match the correct type of rock to how it is formed.

Type of rock
Igneous
Metamorphic
Sedimentary

How the rock is formed
Heat and pressure on existing rocks
Magma cooling
Rock particles being deposited on the sea bed

(2 marks)

8 a) What word is used to describe fractures in layers of rock?

b) What is meant by 'tectonic activity'?

c) What is the Earth's 'lithosphere'? (3 marks)

9 The Earth is made up of a number of plates. These plates move. Explain why the plates move. (3 marks)

Total: 30 marks

PATTERNS OF CHEMICAL CHANGE

Hazard symbols

The hazard symbols below are used to label chemicals to show how dangerous they are.

Oxidising
These substances provide oxygen which allows other materials to burn more fiercely.

Highly flammable
These substances easily catch fire.

Toxic
These substances can cause death. They may have their effects when swallowed, breathed in or absorbed through the skin.

Corrosive
These substances attack and destroy living tissues, including eyes and skin.

Harmful
These substances are similar to toxic substances but are less dangerous.

Irritant
These substances are not corrosive but can cause reddening or blistering of the skin.

Figure 7.1 Hazard warning symbols

Chemical reactions

Here are the main features of a chemical reaction:

1 Chemical reactions make new substances.
2 The starting materials are called **reactants** and the new substances made are called **products**.
3 Before a chemical change can start, the reactants must be mixed so that the particles in them can collide.
4 When the particles collide, the energy of the collision must be enough to break the chemical bonds in the reactants.
5 The least amount of energy needed to do this is called the **activation energy**.
6 Once the reactants have decomposed, the particles released join together in a different way to make the new substances – the products.

Chemical reactions can be represented using different types of equations:

- **General equations** show the types of substances involved in the reaction.
  e.g. acid + alkali → salt + water
- **Word equations** show the **names** of all the reactants and products.
  e.g. hydrochloric acid + sodium hydroxide → sodium chloride + water
- **Symbol (or formula) equations** show the **chemical symbols** or **formulae** of the reactants and products.
  e.g. $HCl + NaOH \rightarrow NaCl + H_2O$

CHECK YOUR UNDERSTANDING

❶ A hazard symbol consisting of a black X can have two meanings. What are they?

❷ What is meant by the term 'activation energy'?

Rates of reaction

Some chemical reactions can be very fast, like explosions, but some can be quite slow, like the fermentation of sugar in grapes to make wine, or the rusting of iron. Figure 7.2 shows some ways of following the rate of chemical reactions.

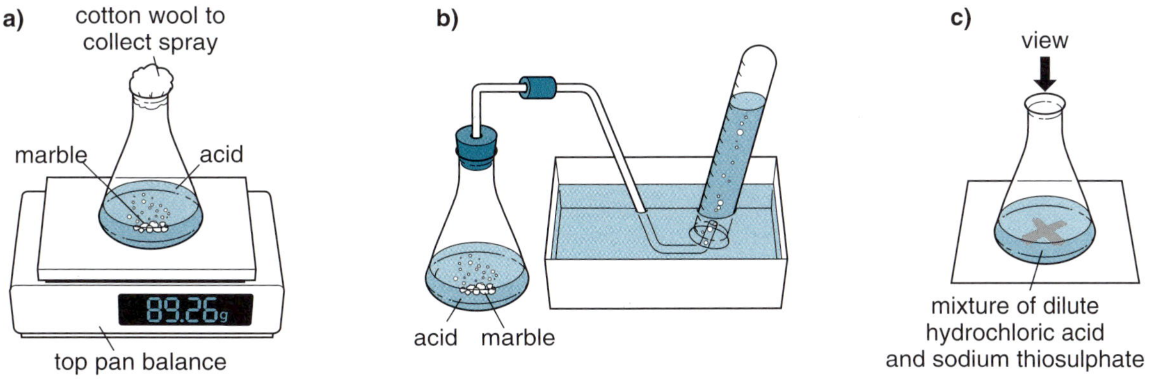

a) Record the mass each minute as the gas escapes.

b) Record the total volume of gas collected each minute.

c) Measure how long it takes for the X to disappear as the liquid goes cloudy.

Figure 7.2 Ways of measuring the rate of reaction

Once a set of readings has been taken, a change is then made (e.g. raising the temperature), and the experiment is repeated.

The readings taken can be displayed on a graph.

The graph for experiment a) above would look something like Figure 7.3.

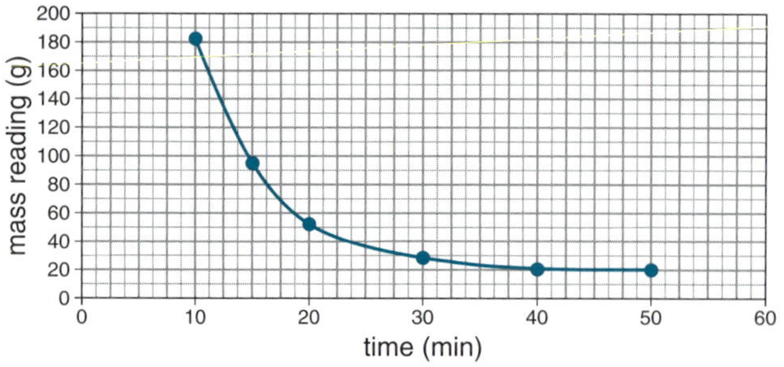

Figure 7.3 A graph of results for experiment a)

Note that as the reaction takes place, the rate slows down – the graph becomes less steep, and eventually it levels off because the reaction is complete. The reaction stops when one of the reactants is used up.

The graph for experiment b) would look something like Figure 7.4.

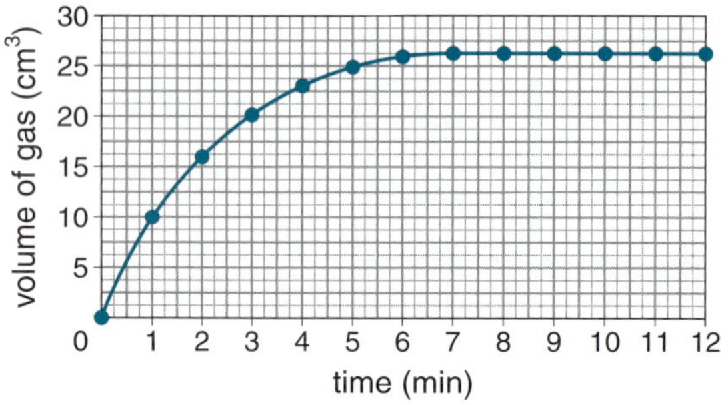

Figure 7.4 A graph of results for experiment b)

Note once again that the graph is quite steep at the beginning and then becomes less steep as the reaction rate slows down. This is because the reactants are slowly being used up and their concentration is decreasing. Eventually no more gas is given off because one of the reactants has been used up.

Changing reaction rates

There are several ways of changing the rate of a reaction. Here is a summary with explanations.

LEARN what can be done to change the reaction rate.

Change	What it does	How it does it
Temperature	Reactions go faster at higher temperatures.	Raising the temperature increases the speed of the reacting particles so that they collide more often and with greater energy so that there is more chance that they will react.
Concentration	Concentrated solutions react faster than dilute solutions.	Increasing the concentration of a solution puts the reactants closer together so that they are likely to collide more often. Also they don't have so far to travel before they can collide.
Pressure	Changing the pressure only affects reactions where gases are reacting. Gas reactions go faster at high pressures.	Increasing the pressure on a gas reaction has exactly the same effect as increasing the concentration of a solution. The reacting particles are closer together and more likely to collide and react.
Particle size	Small particles react faster than large ones.	Smaller particles have a greater surface area than the same mass of larger particles. This means that more solid reactant becomes exposed and there is more chance of a reaction taking place.
Use of a catalyst	Catalysts speed up chemical reactions but do not get used up. They can be used over and over again.	Catalysts provide an easier route from reactants to products. Different reactions usually need different catalysts. Some reactions are not affected by catalysts.

It is important to be able to control the rate of a reaction in industry.

- Too fast a reaction rate could be dangerous and difficult to control.
- Too slow a reaction rate could be costly because it takes too long to make the product.

CHECK YOUR UNDERSTANDING

❶ Which hazard symbol would you put on each of the following: a) a bottle of concentrated sulphuric acid, b) a can of petrol?

❷ When sodium is added to water, sodium hydroxide and hydrogen are made. Write down the name of one reactant and one product in this reaction.

❸ Make a sketch of the graph for experiment a) on page 63 and on it sketch another graph to show what would happen if the experiment was repeated using hot acid instead of cold acid.

Rate of reaction and amount of reaction

The rate of reaction tells you how fast the reaction is going and depends on the factors shown in the table.

The amount of reaction tells you how much reaction is taking place and the amount of product made. This depends on the amounts of reactants used in the reaction. The greater the amount of starting materials, the greater the amount of products formed. The graphs in Figure 7.5 show the differences.

Don't confuse rate of reaction and amount of reaction.

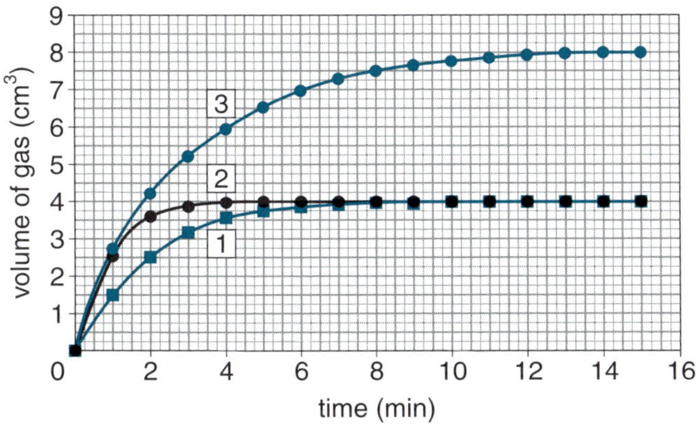

Figure 7.5

Graphs 1 and 2 are for two experiments that have used the same amounts of reactants e.g. 50 cm^3 of acid and a small (1 cm) piece of magnesium. In both experiments there is enough acid to dissolve all the magnesium. So the reaction stops when the magnesium has been completely dissolved.

However, the acid used in experiment 2 is more concentrated than the acid used in experiment 1, so graph 2 is steeper than graph 1. The more concentrated acid is dissolving the magnesium faster but the same total amount of gas is being made in each experiment.

Graph 3 shows what happens if a larger amount of magnesium (e.g. 2 cm) is added to 50 cm^3 of acid and left until it completely dissolves. The greater amount of magnesium helps to make more gas than in the other two experiments.

CHECK YOUR UNDERSTANDING

1. A pupil wants to dissolve a large lump of marble (calcium carbonate) in some hydrochloric acid. Suggest three things the student could do to make the marble dissolve as quickly as possible.
2. Why does a log of wood take longer to burn than when it is chopped up into sticks?
3. Why does food stay fresh for longer in a refrigerator?

Enzymes

Some important points to remember about enzymes:

1 Enzymes are catalysts that are produced by living things.

2 Enzymes are very important because they allow chemical reactions to take place quickly at ordinary temperatures and pressures. This means that the chemical processes will be cheaper to run because there is no need to provide (expensive) heat energy.

3 Enzymes work best in warm conditions rather than hot. This is because they are protein molecules which are usually damaged by temperatures above 45°C.

4 Enzymes are affected by the pH value of their surroundings. Different enzymes work best at different pH values.

Reactions that use enzymes to help them work faster

Making beer and wine 	This process is called fermentation. The enzymes in yeast cells help to convert sugar into alcohol and carbon dioxide.
Making bread dough rise 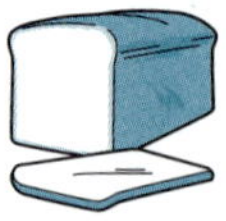	This is also fermentation. The carbon dioxide gas produced helps the bread dough to rise.
Making yoghurt 	Enzymes in bacteria convert the sugar in milk (called lactose) into yoghurt (which contains lactic acid).

Using detergents 	Biological detergents may contain enzymes. These help to digest protein-based stains and fat-based stains. They work at low temperatures.
In some baby foods 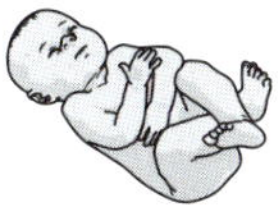	Some baby foods contain enzymes that help to start the process of digesting proteins, making it easier for the baby.
Making sugar syrup 	Starch syrup from plants is converted into a sugar syrup containing the sweet sugars **glucose** and **maltose**. These can be used in the brewing industry and for sweetening drinks.
Making sweeteners 	The enzyme **isomerase** changes glucose into a sweeter sugar called fructose. This means less sugar is needed to sweeten foods and drinks.

CHECK YOUR UNDERSTANDING

❶ **What are enzymes?**

❷ **Why do enzymes stop working at temperatures above about 45°C?**

Energy changes

When chemical reactions take place, energy changes often occur. Sometimes energy is given off in the form of heat and/or light, but sometimes heat energy has to be added to make the reaction take place.

When a fuel (e.g. coal, coke, oil, natural gas or petrol) burns, heat and light energy are released and the surroundings become warmer.

When energy is **released** in this way, the reaction is said to be **exothermic**. These are the most common types of reaction and some other examples are:

- Adding a fairly reactive metal (e.g. magnesium) to an acid.
- Reacting an acid with an alkali (e.g. adding hydrochloric acid to sodium hydroxide).
- Adding water to some white anhydrous copper sulphate powder. This reaction can be used to show that water is present. The water turns the white anhydrous copper sulphate to blue.
- Adding very reactive metals (e.g. sodium and potassium) to water.

Figure 7.6 An exothermic change

When a plant is in sunlight, the process called **photosynthesis** takes place and the plant absorbs energy from the Sun.

When energy is **absorbed** in this way, the reaction is said to be **endothermic**. This type of reaction is not quite so common, but there are other examples. If a chemical compound needs to be heated to decompose it, then an endothermic change is taking place. Here are some more examples:

- Heating copper carbonate to make copper oxide and carbon dioxide.
- Heating blue copper sulphate crystals to make white anhydrous copper sulphate.

Figure 7.7 An endothermic change

The reason why energy is involved in chemical reactions is really quite simple.

1 The chemical bonds holding the elements together in the reactants have to be broken before any new substance can be made.

2 This means heat energy must be added.

3 When new chemical bonds are made in the products, heat energy is given out.

4 If more energy is given out making new bonds than is added to break existing bonds, then the reaction is exothermic. The difference between the energy put in and the energy released is given out, usually as heat energy, so the reaction mixture gets hotter.

5 If less energy is given out making new bonds than has to be added to break existing bonds, then the reaction is endothermic and the reaction mixture gets colder.

These energy changes can be shown using an energy diagram.

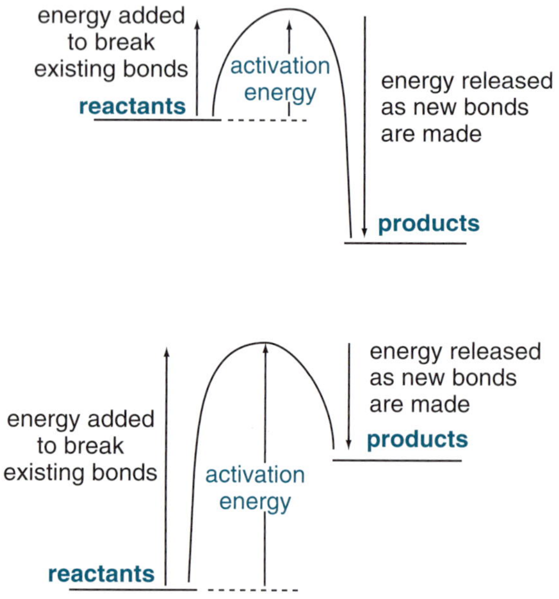

This is an energy diagram for an **exothermic** reaction.

This is an energy diagram for an **endothermic** reaction.

Figure 7.8 Energy diagrams

REMEMBER – The **activation energy** is the least amount of energy needed to start the reaction.

Reversible reactions

Some reactions are called **reversible reactions**, and are shown in chemical equations using a special sign between reactants and products rather than an arrow.

$$\text{Reactants} \rightleftharpoons \text{Products}$$

This sign shows that the reaction can take place both in the forward and the reverse directions depending on the conditions chosen.

One common example is when ammonium chloride solid is heated and the products are cooled:

$$\underset{\text{(a white solid)}}{\text{ammonium chloride}} \underset{\text{cool}}{\overset{\text{heat}}{\rightleftharpoons}} \underset{\text{(colourless gases)}}{\text{ammonia + hydrogen chloride}}$$

REMEMBER – It is the conditions chosen that determine which way the reaction goes.

The forward reaction needs heat energy to make it work so it is endothermic.

The reverse reaction gives out heat energy so it is exothermic.

Another example is the action of heat on blue copper sulphate crystals to make white anhydrous copper sulphate and water. Heat is needed to change the blue crystals into the white powder but heat is released when water is added back to the white powder.

Any reversible reaction that is exothermic in one direction is endothermic in the other.

CHECK YOUR UNDERSTANDING

1. Why is it necessary to put energy into chemical reactions to get them started?
2. What energy change takes place when new chemical bonds are made?
3. What type of change (exothermic or endothermic) takes place when calcium carbonate is heated to form calcium oxide and carbon dioxide?

Some important industrial processes

The air around us is a mixture of gases including nitrogen, oxygen, noble gases, carbon dioxide and water vapour. About 80% of the air is nitrogen gas and this is an essential element needed to make plants grow. Unfortunately few plants can use the nitrogen directly from the air and it is often necessary to add extra nitrogen to crops in the form of **nitrogen-based fertilisers**. These are manufactured from a compound called ammonia (NH_3) which contains nitrogen. Here are some advantages and disadvantages of using artificial fertilisers.

Advantages	Disadvantages
1 Crop yields are improved.	1 If too much fertiliser is added, some gets washed off the land into streams and rivers and the water becomes contaminated.
2 The land used for growing crops can be used every year for growing more crops instead of leaving it every so often to recover.	2 The fertiliser in the water encourages water plants and algae to grow. The surface becomes covered in green algae which are decomposed by bacteria using up oxygen dissolved in the water. Fish are then starved of essential oxygen.
	3 The nitrates in fertilisers get into drinking water and these nitrates are converted by the body into harmful chemicals.

The Haber process for making ammonia

Ammonia (NH_3) is manufactured from nitrogen and hydrogen in the **Haber process**.

The reaction is a reversible one.

$$\text{nitrogen} + \text{hydrogen} \rightleftharpoons \text{ammonia}$$
$$N_2 + 3H_2 \rightleftharpoons 2NH_3$$

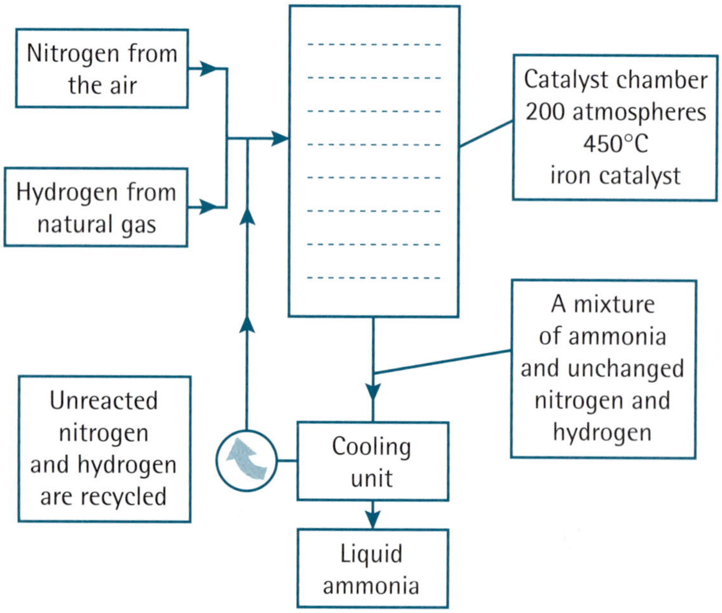

Figure 7.9 The Haber process

Only small amounts of the nitrogen and hydrogen in the mixture actually combine. The mixture leaving the chamber is cooled and the ammonia gas condenses to liquid ammonia. The unchanged nitrogen and hydrogen are then recycled back through the catalyst chamber.

The important points to learn about the Haber process are summarised below.

Feature	What you need to know
What are the raw materials?	Nitrogen from the air and hydrogen obtained from natural gas.
How are they combined?	They are first purified and are then passed over a catalyst at a moderate temperature and a high pressure.
What pressure is used and why?	A high pressure of about 200 atmospheres is used. High pressure increases the rate of this reaction and helps to make more ammonia.
What temperature is used and why?	A moderate temperature of about 450°C is used. A very high temperature would speed up the reaction but it would also make the ammonia decompose back into nitrogen and hydrogen. Low temperatures make the reaction too slow.
Which catalyst is used?	A catalyst of iron is used.

LEARN these important points about the Haber process.

Changing the ammonia into fertiliser

Stage	What happens
1	A mixture of ammonia and air is passed over a platinum catalyst at 900°C. The ammonia is oxidised to nitrogen monoxide (NO) and water (H_2O).
2	The nitrogen monoxide is cooled and reacted with more oxygen to make nitrogen dioxide (NO_2).
3	The nitrogen dioxide then reacts with oxygen and water to make nitric acid (HNO_3).
4	The nitric acid can then be reacted with ammonia to make the fertiliser called ammonium nitrate which has the chemical formula NH_4NO_3.

Ammonia can be reacted with sulphuric acid to make another fertiliser called ammonium sulphate.

CHECK YOUR UNDERSTANDING

1. What is meant by a reversible reaction?
2. Describe a simple test you could do to show that a liquid contains water. Give the result of the test.
3. Make a list of the important reaction conditions used in the Haber process.

Simple calculations

Every substance is made from **elements** and each element is made of **atoms**. Atoms of different elements have **different masses**. The masses of individual atoms are too small for us to consider so we need to know how the masses of different elements compare with each other. This is done by using **relative atomic masses** (the symbol for which is A_r, see page 43).

Relative atomic mass (A_r)

H = 1 C = 12 O = 16 Na = 23 Mg = 24 S = 32 Cl = 35.5

These numbers **compare** the masses of atoms of different elements with each other. For example, Mg = 24 and C = 12 tells us that a magnesium atom is twice as heavy as a carbon atom. C = 12 and H = 1 tells us that carbon atoms are twelve times heavier than hydrogen atoms. When we weigh out chemicals we have to allow for these differences in masses.

REMEMBER – Relative atomic masses are always given on the Data Sheet issued with the exam paper.

Relative formula mass (M_r)

Atoms are rarely found on their own, they are nearly always combined together. Sometimes identical atoms join together to make **molecules** of elements, e.g. H_2. Sometimes atoms of different elements join together to make **compounds**, e.g. H_2SO_4. To find out how the masses of these substances compare, we must use **relative formula masses** (the symbol for which is M_r).

Relative formula masses are calculated by adding together the masses of all the atoms of each element shown in the chemical formula.

Example 1: What is the relative formula mass of carbon dioxide (CO_2)?

C = 12 O = 16

The formula (CO_2) tells us that there is **1** carbon atom and **2** oxygen atoms.

So: CO_2 has a relative formula mass of 12 + 16 + 16 = 44.

Example 2: What is the relative formula mass of sulphuric acid (H_2SO_4)?

H = 1 S = 32 O = 16

The formula (H_2SO_4) tells us that there are **2** hydrogen atoms, **1** sulphur atom and **4** oxygen atoms.

So: H_2SO_4 has a relative formula mass of (1 + 1) + 32 + (16 + 16 + 16 + 16) = 98.

To calculate the percentage of an element in a compound follow these steps:

1 Calculate the relative formula mass of the compound as shown above.
2 Calculate the fraction of the element present in the compound.
3 Convert this to a percentage by multiplying by 100.

Example: What is the percentage of magnesium (Mg) in magnesium sulphate ($MgSO_4$)?

1 The relative formula mass of $MgSO_4$ is 24 + 32 + 16 + 16 + 16 + 16 = 120.
2 The fraction of magnesium in this compound is 24 ÷ 120.
3 The percentage of magnesium is 24 ÷ 120 × 100 = 20%.

CHECK YOUR UNDERSTANDING

❶ The relative atomic masses of oxygen and sulphur are as follows: O = 16, S = 32. What does this tell you about the masses of sulphur and oxygen atoms?

❷ A student worked out the relative formula mass of sulphur dioxide (SO_2) as 48. Was the student correct? (O = 16, S = 32)

❸ What is the total number of atoms in one molecule of sulphuric acid (H_2SO_4)?

Topic Test

1 What does this hazard symbol show? (1 mark)

2 One of the hazard symbols shows a skull and crossbones. What does this mean? (1 mark)

3 Use words from the following list to copy and complete the sentences below.

activation broken collide new products reactants

In a chemical reaction, the starting materials, called __________, must first __________ with each other so that the bonds in them can be __________.

The least amount of energy needed for this to take place is called the __________ energy.

The starting materials change into __________ substances called __________. (6 marks)

4 Explain why magnesium dissolves faster in concentrated acid than in dilute acid. (2 marks)

5 What is a catalyst and what does it do to the rate of a chemical reaction? (2 marks)

6 Why do most chemical reactions become slower as the reaction proceeds? (3 marks)

7 Why do chemical reactions eventually stop? (1 mark)

8 Why are many industrial chemical reactions involving gases carried out at quite high pressures? (2 marks)

9 What is an enzyme? (1 mark)

10 Name three processes that use enzymes. (3 marks)

11 What are the main advantages of using enzymes in industry? (1 mark)

12 Some reactions are described as being exothermic and some as endothermic. Explain the terms exothermic and endothermic. (2 marks)

13 Give one example of an exothermic reaction and one example of an endothermic reaction. (2 marks)

14 What name is given to the minimum amount of energy that must be given to the reacting substances to start a chemical reaction? (1 mark)

15 Describe a simple test for the gas carbon dioxide. (2 marks)

16 What is meant by a reversible reaction? (1 mark)

17 What special sign is used to show that a reaction is reversible? (1 mark)

18 Give one advantage and one disadvantage of using fertilisers. (2 marks)

19 Ammonia is made industrially from nitrogen and hydrogen. Where are these two gases obtained from for this process. (2 marks)

20 List the essential conditions of temperature and pressure used in the Haber process for manufacturing ammonia. (2 marks)

21 Which catalyst is used in the Haber process? (1 mark)

22 Why is a very high temperature not used in the Haber process? (1 mark)

23 Give the name of a common fertiliser made from ammonia. (1 mark)

24 The formulae of two fertilisers are a) NH_4NO_3 and b) $(NH_4)_2SO_4$. For each fertiliser calculate the relative formula mass and the percentage of nitrogen present. (Use the following relative atomic masses: H = 1; N = 14; O = 16; S = 32.) (4 marks)

Total: 45 marks

STRUCTURES AND BONDING

The atom

Some important points about atoms:

- All substances are made from atoms.
- If all the atoms in a substance are the same, the substance is an element.
- If the atoms in a substance are different and are joined together, the substance is a compound. (If they are not joined together, it is a mixture.)
- Atoms have a small central part called the nucleus. This contains positively charged protons, together with neutrons that have no electrical charge.
- All atoms of a particular element have the same number of protons.
- Atoms of different elements have different numbers of protons.
- Atoms of the same element can have different numbers of neutrons. These atoms are called isotopes.
- Nearly all the mass of the atom is in its nucleus.
- The nucleus is surrounded by a series of shells or energy levels. These contain negatively charged electrons.
- Electrons have almost no mass.
- In an atom, the number of protons in the nucleus equals the number of electrons in the energy levels. Atoms have no overall electrical charge.
- The number of protons in an atom is called the atomic number.
- The total number of protons and neutrons in an atom is called the mass number.

LEARN these important points about atoms.

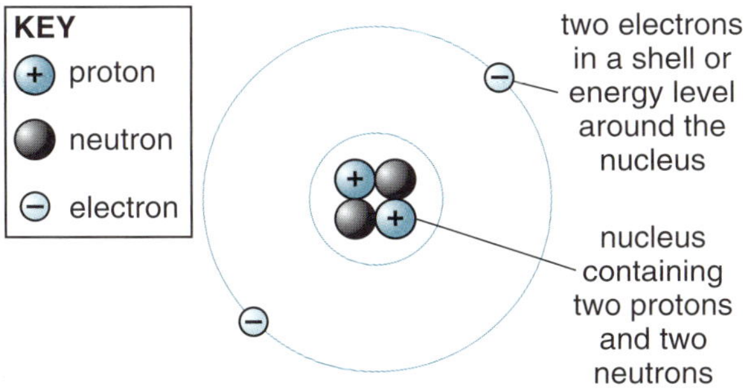

Figure 8.1 An atom of helium

The relative masses and charges of protons, neutrons and electrons:

	Mass	Charge
Proton	1	+1
Neutron	1	0
Electron	negligible	−1

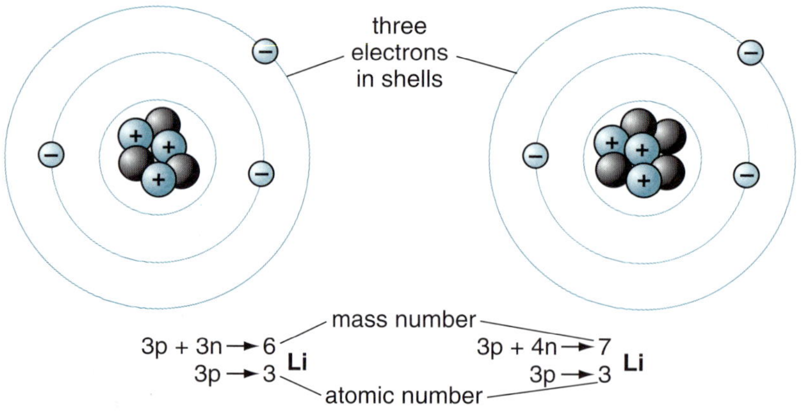

Figure 8.2 Isotopes of lithium – element number 3

> **REMEMBER** – **Isotopes** are atoms of the **same** element with the **same number** of **protons** but **different numbers** of **neutrons**.
>
> **Atomic numbers** and **mass numbers** are shown as follows:
> for sodium –
>
> mass number 23
> Na
> atomic number 11

The outer part of the atom

Each electron occupies a particular energy level (or shell). Where there are several energy levels available, the electron occupies the level with the lowest energy.

The electronic structures of elements can be shown in the form of a diagram (see Figure 8.3) or a table.

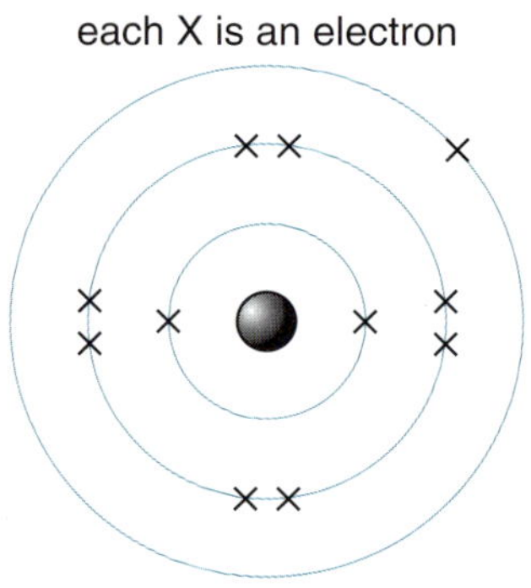

Figure 8.3 An atom of sodium 2,8,1

> Some important points to note:
> 1 The first energy level (the one closest to the nucleus) can only hold a maximum of **2** electrons.
> 2 The second shell can hold a maximum of **8** electrons.
> 3 The third shell is stable when it has **8** electrons in it.
> 4 When the shells in an atom are full, the atom is stable.
> 5 The electronic structure shown in Figure 8.3 (2,8,1) shows the numbers of electrons in each energy level starting with the ones closest to the nucleus.

The electronic structures of the first 20 elements in the Periodic Table are shown in the table below.

Atomic number	Name	Symbol	1st energy level	2nd energy level	3rd energy level	4th energy level
1	hydrogen	H	1			
2	helium	He	2	1st energy level now full		
3	lithium	Li	2	1		
4	beryllium	Be	2	2		
5	boron	B	2	3		
6	carbon	C	2	4		
7	nitrogen	N	2	5		
8	oxygen	O	2	6		
9	fluorine	F	2	7		
10	neon	Ne	2	8	2nd energy level full	
11	sodium	Na	2	8	1	
12	magnesium	Mg	2	8	2	
13	aluminium	Al	2	8	3	
14	silicon	Si	2	8	4	
15	phosphorus	P	2	8	5	
16	sulphur	S	2	8	6	
17	chlorine	Cl	2	8	7	
18	argon	Ar	2	8	8	3rd level full
19	potassium	K	2	8	8	1
20	calcium	Ca	2	8	8	2

CHECK YOUR UNDERSTANDING

1. Using an x to represent each electron, draw diagrams similar to Figure 8.3 to show how the electrons are arranged in atoms of a) silicon and b) argon.
2. How many protons, neutrons and electrons are there in an atom of the isotope of chlorine that has an atomic number of 17 and a mass number of 35?

Electronic structures give us lots of important information. Check the points on the next page by looking at the table above and a copy of the Periodic Table.

1. The **noble gas** elements have all their energy levels **full**.
2. The number of electrons in the outermost energy level tells us the **group number** to which the element belongs. For example, sodium (2,8,1) is in Group 1; fluorine (2,7) is in Group 7. The group number is the number at the top of each vertical column in the Periodic Table.
3. The rows in the Periodic Table are known as **periods**. The period number tells us the number of the energy level being filled. For example, magnesium is in the third period because the third energy level is being filled.
4. Metal elements usually have one, two or three electrons in their outer energy level.
5. Non-metal elements usually have four, five, six or seven electrons in their outer energy level.
6. Noble gases usually have eight electrons in their outer energy level (except helium that has only two).
7. It is also important to know that when atoms combine, they do so to obtain the same electronic structure as the nearest noble gas (see pages 79–82).

Organising elements

Early attempts at arranging the elements into order made use of relative atomic masses (A_r) (see page 72). The elements were arranged in order of their relative atomic masses, and for most elements this placed them in groups with very similar chemical properties. This had some problems, for example argon atoms have a greater relative atomic mass than potassium atoms but argon is better placed before potassium in the Periodic Table. The modern Periodic Table arranges the elements in order of their atomic number (i.e. the number of protons in the nucleus). Arranging them this way puts each element into its correct group. Note that there are more metals than non-metals.

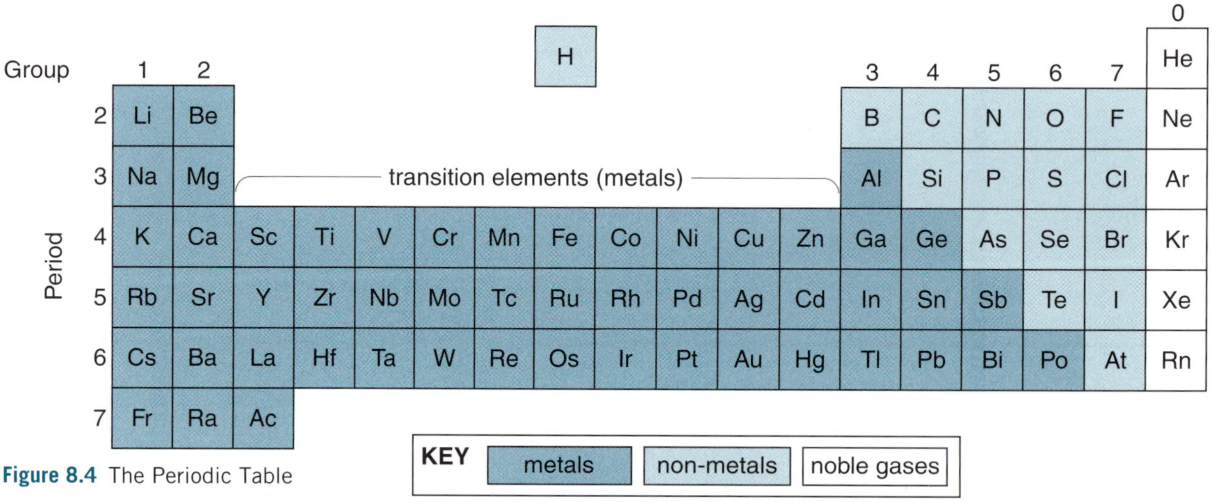

Figure 8.4 The Periodic Table

Ionic bonding

The main features of ionic bonding are:

1. Ionic bonds are formed between metals and non-metals.
2. Electron(s) transfer from the outer energy level (shell) of the metal atoms to the outer energy level (shell) of the non-metal atoms.
3. Once the electrons have transferred, the atoms have changed into positively charged metal ions and negatively charged non-metal ions usually with noble gas structures.
4. The oppositely charged ions are strongly attracted together by electrostatic forces of attraction and an ionic bond is made.

REMEMBER – In ionic bonding electrons transfer from metal to non-metal.

Here is an example:
sodium ($_{11}Na$) joining with chlorine ($_{17}Cl$) to make sodium chloride (NaCl).

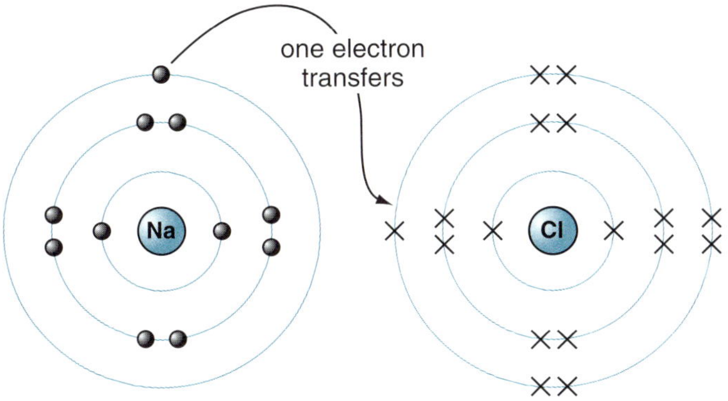

Sodium has one electron in its outer energy level.
If it loses it, the electron structure will be the same as neon.

Chlorine is short of one electron in its outer energy level.
If it gains one, the electron structure will be the same as argon.

One electron transfers from sodium (Na) to chlorine (Cl) making Na^+ and Cl^- ions.

The Na^+ ion has a single positive charge because it has **lost** a negatively charged electron.

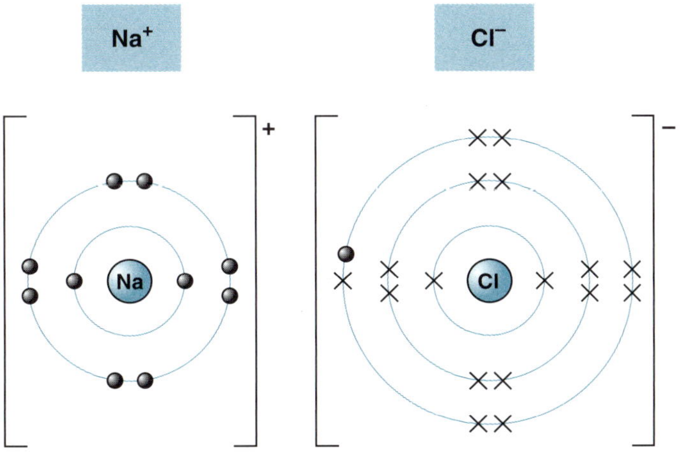

The Cl^- ion has a single negative charge because it has **gained** a negatively charged electron.

The two oppositely charged ions are **electrostatically attracted** and the compound sodium chloride (NaCl), containing an **ionic bond**, is made.

Figure 8.5 Ionic bonding in sodium chloride

Sometimes more than one electron is transferred from the metal atom to the non-metal atom forming ions that have more than one positive or negative charge.

Sometimes more than one atom of metal or non-metal is needed.

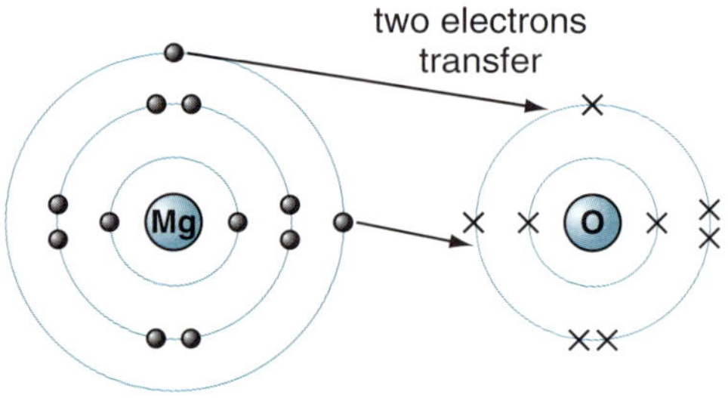

Figure 8.6 Ionic bonding in magnesium oxide

This makes Mg^{2+} and O^{2-} ions which attract to make magnesium oxide (MgO).

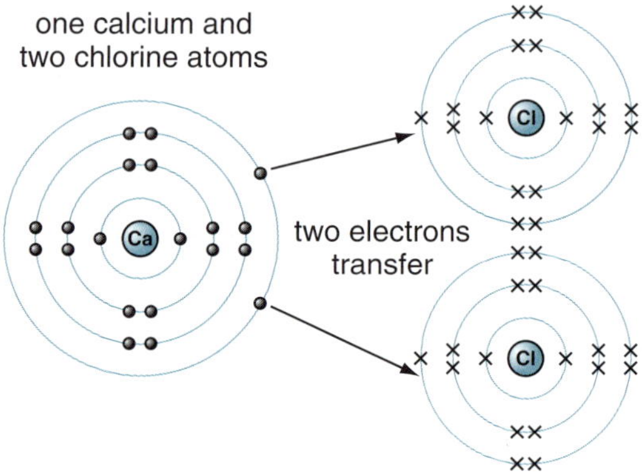

Figure 8.7 Ionic bonding in calcium chloride

This makes one Ca^{2+} and two Cl^- ions which attract to make calcium chloride ($CaCl_2$).

Can you work out how sodium (Na) and oxygen (O) combine to make sodium oxide (Na_2O)?

Extra information:

1 Metals have one, two or three electrons in their outer energy level (shell) and they always **transfer** all of them when making ionic bonds.
2 Non-metals are short of electrons in their outer energy level (shell) and they **gain** as many electrons as are needed to fill the outer energy level.
3 The direction of electron transfer is always **from metal to non-metal**.
4 **Metals** always form **positive ions**, **non-metals** always form **negative ions**.
5 The strong electrostatic forces of attraction between lots of ions holds the ions tightly together so ionic compounds are **solids** with **high melting points**.
6 The nucleus containing protons and neutrons does not change as the bond is made.
7 Solid ionic compounds do not conduct electricity because the ions are not free to move and carry the current. Melting the solid or dissolving it in water allows the ions to move and conduct electricity. This is called **electrolysis** (see page 47).

CHECK YOUR UNDERSTANDING

❶ Write down the three main stages in the formation of an ionic bond.

❷ Draw 'dot and cross' diagrams to show how lithium ($_3Li$) and oxygen ($_8O$) could form the ionic compound lithium oxide (Li_2O). Include the charges on the ions.

Covalent bonding

REMEMBER – In covalent bonding the electrons are shared in pairs.

The main features of covalent bonding are:

1 Covalent bonds are formed between non-metal atoms.

2 Electrons are shared in pairs between the atoms, each atom donating one electron to each shared pair.

3 Each atom shares as many of its electrons as are needed to allow its outer energy level to become full.

4 Once the electrons have been shared, the particles formed are usually small separate molecules.

5 The forces holding the atoms together in each molecule are strong but often the forces between adjacent molecules are quite weak so covalent substances are either gases, low boiling-point liquids or low melting-point solids.

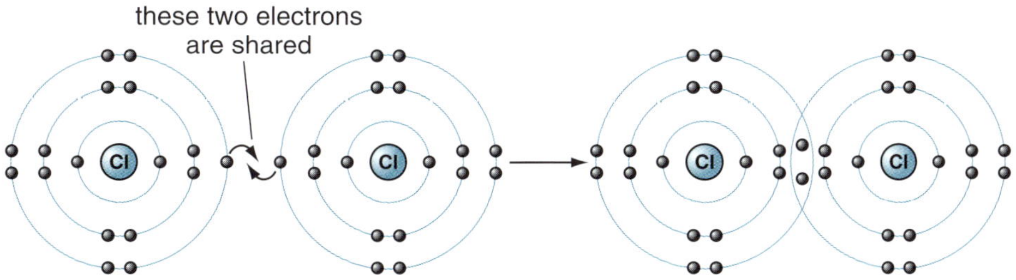

Figure 8.8 Covalent bonding in a chlorine molecule

In Figure 8.8, two chlorine atoms (Cl) have donated one electron each to a shared pair of electrons. Note how the two outer energy levels are overlapped to show sharing.

Sometimes an atom of one element has to share its electrons with more than one atom of a different element. This can be seen in water (H_2O), ammonia (NH_3) and methane (CH_4).

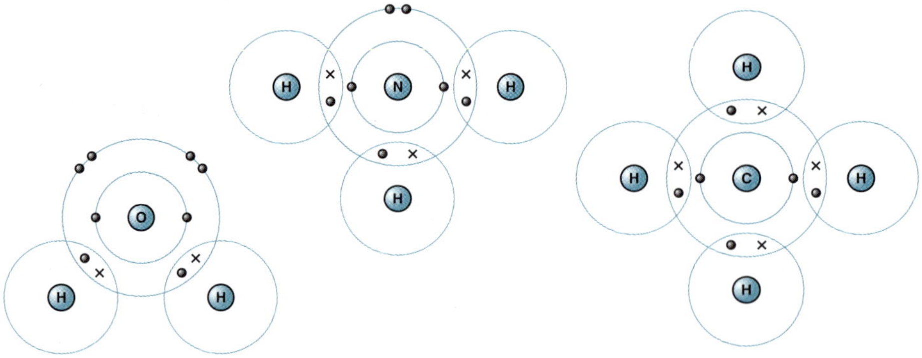

In water (H_2O), two separate covalent bonds are formed with two separate hydrogen atoms.

In ammonia (NH_3), three separate covalent bonds are formed with three separate hydrogen atoms.

In methane (CH_4), four covalent bonds are formed.

Figure 8.9 More examples of covalent bonding

Note that every atom in the structures in Figure 8.9 now has a noble gas structure with all energy levels full.

Sometimes atoms can form **double covalent bonds** with each other. For example in oxygen gas (O_2):

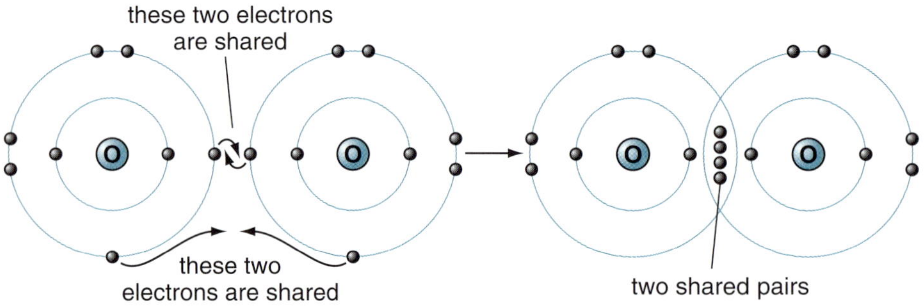

Figure 8.10 Covalent bonding in an oxygen molecule

Another way to show covalent bonds is to use a single line to represent each shared pair of electrons. The examples already shown would become:

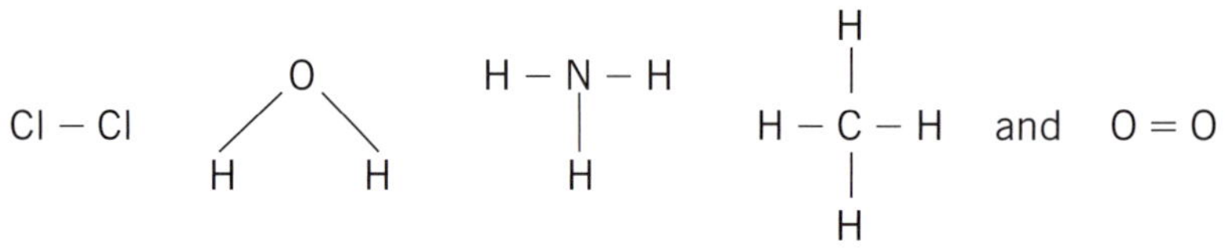

Metallic bonding

The main features of metallic bonding are:

1 Metallic bonds only involve **metal** elements.

2 Metal elements have 1, 2 or 3 electrons in their outer shell and when metal atoms form a piece of metal, these outer electrons are **pooled** to form a **sea of negative charge** which holds together the rest of each atom (the nucleus and the full inner energy levels).

3 These pooled electrons are **free to move** through the metallic structure and allow the atoms to slide over each other (so they can be bent) and also allow them to **conduct electricity and heat** very quickly.

REMEMBER – In metallic bonding the outer electrons are pooled together.

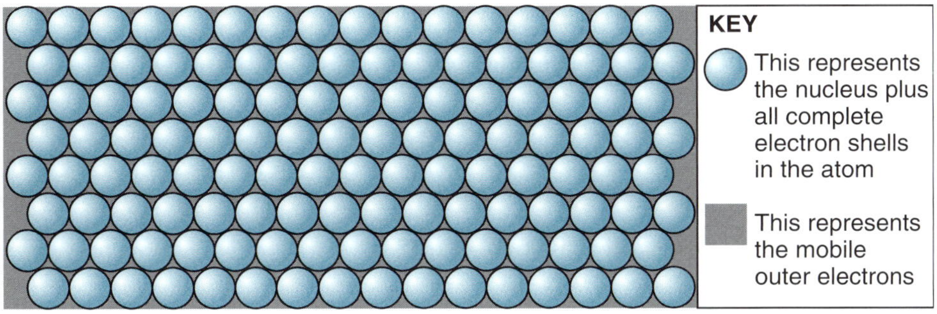

Figure 8.11 The arrangement of atoms and outer electrons in a metal

CHECK YOUR UNDERSTANDING

1. Draw diagrams to show how covalent bonds could be formed between
   a) two hydrogen atoms to make hydrogen gas (H_2)
   b) hydrogen and chlorine to make hydrogen chloride (HCl).
2. Give the name of the type of chemical bond formed between the following pairs of elements:
   a) sulphur and oxygen
   b) copper and chlorine
   c) silicon and hydrogen.
3. What would you expect the physical states of the compounds in Q2 to be (solid, liquid or gas)?
4. Why don't covalent substances conduct electricity?

Variations and similarities within groups

Group 1 – The alkali metals (lithium, sodium and potassium)

What they have in common	Trends down the group
They all have one outer electron which they can lose to form a 1+ ion e.g. Na^+.	The size of the atoms increases because more energy levels (shells) have been added.
They all float on the surface of water and react very vigorously with cold water producing hydrogen and an alkali e.g. sodium hydroxide (NaOH).	The outer electron is further away from the attraction of the nucleus and becomes easier to remove. This means K^+ ions will be more easily formed than Li^+ ions.
They are fairly soft, low density metals and are easily cut.	Reactivity with water and other substances increases.
They are good conductors of heat and electricity.	Melting and boiling points decrease.

Group 7 – The halogens (chlorine, bromine and iodine)

What they have in common	Trends down the group
They all have seven outer electrons and gain one more to form a 1- ion e.g. Cl^-, Br^-, I^-.	The size of the atoms increases because more energy levels (shells) have been added.
They exist as diatomic molecules i.e. molecules composed of two atoms e.g. Cl_2, Br_2, I_2.	Reactivity decreases because an incoming electron cannot get so close to the nucleus.
They all have coloured vapours (Cl_2 is green, Br_2 is red/brown, I_2 is purple).	Melting and boiling points increase down the group.
They form ionic solids when reacted with metals.	There is a change from gas (Cl_2) to liquid (Br_2) to solid (I_2).
They form molecular compounds when reacted with non-metals.	A more reactive halogen (e.g. Cl_2) can displace a less reactive halogen (e.g. Br_2 or I_2) from a solution of its salt (e.g. KBr or KI).

Group 0 – The noble gases (helium, neon, argon)

What they have in common	Trends down the group
They all have full outer energy levels.	The size of the atoms increases because more energy levels (shells) have been added.
They are chemically unreactive.	
They exist as individual atoms rather than diatomic molecules like other gases.	
They are used as inert gases in light bulbs and discharge tubes.	

CHECK YOUR UNDERSTANDING

1. What are the vertical columns in the Periodic Table called?
2. How does reactivity of the elements change as you go down Group 1?
3. Why are helium, neon and argon unreactive elements?

An important use for salt

Common salt is **sodium chloride** (NaCl). A solution of salt is called **brine** and this solution can be decomposed using electricity in a process called **electrolysis** (see page 47).

Sodium chloride solution contains four ions: Na^+ ; H^+ ; OH^- and Cl^-.

The positively charged **anode** attracts the OH^- and Cl^- ions.
The Cl^- ions lose electrons to the anode and become chlorine atoms (Cl).
These chlorine atoms join together to make chlorine molecules (Cl_2).
Chlorine gas is made at the anode.

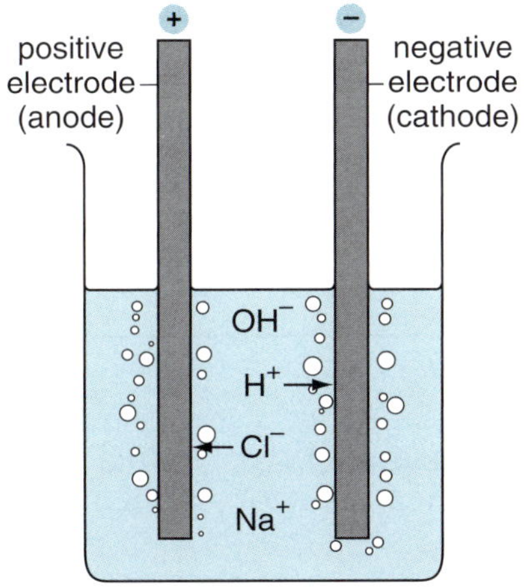

The negatively charged **cathode** attracts the Na^+ and H^+ ions.
The H^+ ions gain electrons from the cathode and become hydrogen atoms (H).
These hydrogen atoms join together to make hydrogen molecules (H_2).
Hydrogen gas is made at the cathode.

A test for chlorine gas is that it **bleaches damp litmus paper**.

The Na^+ and OH^- ions stay in solution forming the alkali sodium hydroxide (NaOH).

A test for hydrogen gas is that it **burns with a squeaky explosion**.

Figure 8.12 The electrolysis of sodium chloride solution

Uses for the substances made from salt

Substance	Uses
chlorine	Killing bacteria in drinking water and swimming baths. Manufacturing hydrochloric acid, disinfectants, bleach and the plastic PVC.
hydrogen	Manufacturing ammonia and margarine.
sodium hydroxide	Manufacturing soap, paper and ceramics.

Silver halides and hydrogen halides

Silver chloride (AgCl), silver bromide (AgBr) and silver iodide (AgI) are **silver halides** which can be **reduced** to silver (Ag) by the action of **light**, **X-rays** and the **radiation** from radioactive substances.

e.g. $2AgBr \rightarrow 2Ag + Br_2$

They are used to make **photographic film** and **photographic paper**.

Hydrogen chloride (HCl), hydrogen bromide (HBr) and hydrogen iodide (HI) are **hydrogen halides**.

They are gases that dissolve in water to make acidic solutions, e.g. hydrochloric acid.

CHECK YOUR UNDERSTANDING

1. What name is given to the process in which a chemical substance is decomposed by electricity?
2. When electricity is passed through sodium chloride solution, an alkali is made. What test could you carry out to prove this?

Representing chemical substances in chemical reactions

Word equations

These show the names of the starting materials (**reactants**) and new materials formed (**products**) in a chemical reaction. Some important ones to know are:

- acid + metal → a salt + hydrogen
- acid + metal oxide → a salt + water
  (metal oxides are known as bases)
- acid + alkali → a salt + water
  (an alkali is a soluble base)
- acid + metal carbonate → a salt + water + carbon dioxide

LEARN these four basic word equations.

Chemical symbols for elements and ions

Each element is given a different **symbol**. For example, carbon is C; sodium is Na.

Each ion is also given a symbol. For example, the sodium ion is Na^+; the carbonate ion is CO_3^{2-}.

These symbols can be found on the Data Sheet provided with the examination paper.

Chemical formulae of substances

Each substance is given a **chemical formula** which shows the number of each type of atom present in the substance.

For example, carbon dioxide (CO_2) has one carbon atom (C) and two oxygen atoms (O_2).

Sulphuric acid (H_2SO_4) has two hydrogen atoms (H_2), one sulphur atom (S) and four oxygen atoms (O_4).

Elements		Compounds	
hydrogen	H_2	hydrochloric acid	HCl
nitrogen	N_2	sulphuric acid	H_2SO_4
oxygen	O_2	nitric acid	HNO_3
chlorine	Cl_2	water	H_2O
bromine	Br_2	ammonia	NH_3
ozone	O_3	methane	CH_4
helium	He	sulphur dioxide	SO_2
argon	Ar	carbon dioxide	CO_2

REMEMBER these important formulae.

The formulae of ionic compounds can be worked out by using the symbols for the ions shown on the Data Sheet. (All you do is use enough of each ion to cancel out the electrical charges.)

e.g. sodium chloride is NaCl (i.e. $Na^+ + Cl^-$)
copper sulphate is $CuSO_4$ (i.e. $Cu^{2+} + SO_4^{2-}$)
potassium carbonate is K_2CO_3 (i.e. $2 \times K^+ + CO_3^{2-}$)

Writing balanced chemical equations

Each equation must have the same number of each type of atom on each side.

The physical state of each substance can also be shown. For example:

$$MgCO_3(s) + 2HCl(aq) \rightarrow MgCl_2(aq) + H_2O(l) + CO_2(g)$$

Note here that two HCl are needed to 'balance' the equation.

(s) = solid
(l) = liquid
(g) = gas
(aq) = aqueous solution – a solution in water.

Topic Test

1 Name the small central part of an atom. (1 mark)

2 What name is given to the regions around the central part of the atom where electrons can be found? (1 mark)

3 Give the names, symbols and charges of the two types of particle found in the small central part of an atom. (6 marks)

4 Most elements exist in nature as a mixture of 'isotopes'. What are isotopes? (2 marks)

5 How does a sodium atom change into a sodium (Na^+) ion? (1 mark)

6 How does an oxygen atom change into an oxide (O^{2-}) ion? (1 mark)

7 Copy the following table and fill in the blanks.

Particle	Number of			Atomic number	Mass number
	protons	neutrons	electrons		
N atom	7				14
B atom		6		5	
Al atom		14	13		
Ca^{2+} ion	20				40
F^- ion		10		9	

(15 marks)

8 a) Draw dot and cross diagrams to suggest how magnesium (Mg) and fluorine (F) could form magnesium fluoride (MgF_2).
b) Name the type of bond formed between magnesium and fluorine.
c) What is the likely physical state of magnesium fluoride? (5 marks)

9 The electronic structure of phosphorus (P) is 2,8,5. Without looking at the Periodic Table, suggest, with reasons, the group and period to which phosphorus belongs. (2 marks)

10 Why can all metals conduct an electric current? (1 mark)

11 What trend in reactivity is seen in a) Group 1 and b) Group 7? (2 marks)

12 Why are noble gases stable and unreactive? (1 mark)

13 Name the three new substances formed when an electric current is passed through a solution of sodium chloride in water. (3 marks)

14 What would you see if some white silver bromide was left exposed for some time to bright sunlight? (1 mark)

15 Describe chemical tests, with their results, for the following: a) hydrogen gas, b) chlorine gas. (2 marks)

16 Give one use for chlorine gas. (1 mark)

17 The formulae of some ions are as follows: Na^+; O^{2-}; Mg^{2+}; Br^-. What are the chemical formulae of a) sodium bromide, b) magnesium bromide, c) sodium oxide? (3 marks)

18 How many atoms of each element are there in one molecule of a) nitric acid (HNO_3) and b) ammonium sulphate ($(NH_4)_2SO_4$)? (2 marks)

19 a) Balance the chemical equation $CH_4(g) + \_\_\_\_ O_2(g) \rightarrow CO_2(g) + \_\_\_\_ H_2O(l)$
b) What do the symbols (g) and (l) indicate? (4 marks)

Total: 54 marks

Chemistry examination questions and model answers

1 a) Sodium chloride is an ionic compound.

This is a diagram of a sodium ion.

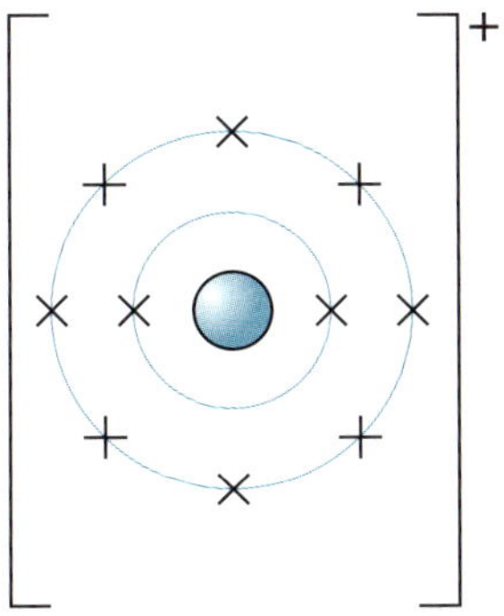

Copy and complete this diagram of a chloride ion.

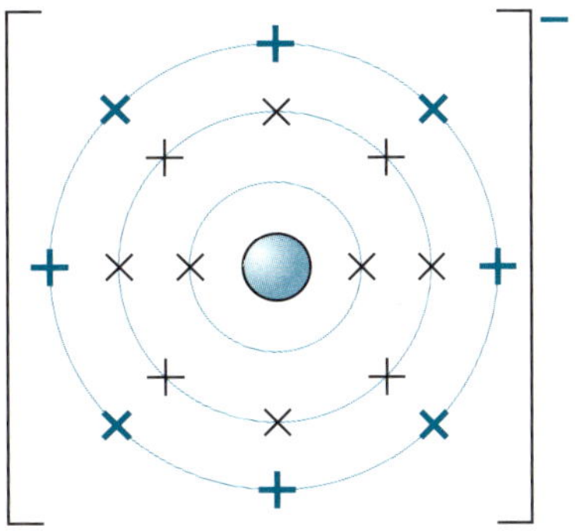

(2 marks)

The chloride ion, like all ions, has a full outer shell. So there must be 8 electrons on the outer ring in the diagram.

Looking at the Periodic Table on the Data Sheet shows that chlorine is in group 7. This means it must have 7 electrons in its outer shell. The chloride ion has 8 electrons in the outer shell and so must have an overall negative charge.

b) The electrolysis of sodium chloride produces chlorine.

Chlorine is used to produce bleach (HOCl).
This equation represents the reaction.

$$Cl_2(g) + H_2O(l) \rightarrow HCl(aq) + HOCl(aq)$$

Give the meaning of the state symbols (l) and (aq).

(l) liquid

(aq) aqueous *(2 marks)*

(l) means the substance is liquid.

(aq) means 'aqueous' but 'dissolved in water' or 'in solution in water' are acceptable answers. The key word here is 'water'. Just saying 'in solution' is not enough for the mark. You will not get credit for words that are like 'aqueous' (for example 'aqua' or 'aquatic').

c) The equation below represents the reaction between chlorine and potassium bromide.

$$Cl_2(aq) + 2KBr \rightarrow 2KCl(aq) + Br_2(aq)$$

(i) Balance this equation. *(1 mark)*

The equation shown is correctly balanced. But $\frac{1}{2}Cl_2 + KBr \rightarrow KCl + \frac{1}{2}Br_2$ is acceptable.

(ii) Why does chlorine displace bromine from potassium bromide solution? *(1 mark)*

Because chlorine is more reactive than bromine.

> The key word here is 'reactive'. You won't get the mark for words like 'stronger'. Saying 'higher in the group' is not an acceptable answer because it does not explain *why* bromine is displaced.
>
> You will also lose the mark if you write 'bromide' instead of 'bromine'.

2 The elements in Group 1 are known as the alkali metals.

Which **three** of the following are properties of alkali metals?

Place a tick (✔) in the box against each correct property. *(3 marks)*

Hard, tough and strong	☐
Low density	✔
Form hydroxides that dissolve in water	✔
React quickly with water	✔
Used as catalysts	☐
Used to make electric cables	☐

> There are no alternative answers in this question. Three answers (ticks) are required. You cannot guarantee getting the right answers by ticking all 6 boxes! As a rule you will limit your maximum mark if you tick too many boxes. For example if 4 boxes are ticked your maximum mark could be reduced to 2.

ENERGY

Transferring heat (thermal) energy

Heat (thermal) energy is transferred from a place at higher temperature to a place at lower temperature by three methods: **conduction**, **convection** and **radiation**. Only radiation allows the transfer of energy through a vacuum. This method does not involve particles of matter.

Radiation

Emitting radiation

Hot objects give out mainly infrared radiation. The hotter the object, the more energy it radiates.

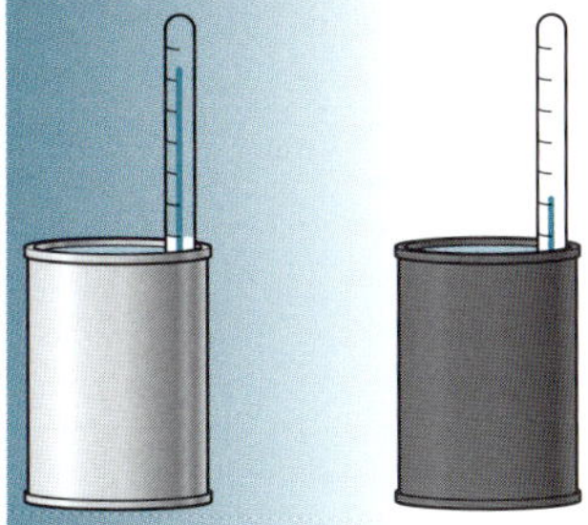

Dark, matt surfaces give out more radiation than light, shiny surfaces at the same temperature.

The dark coloured tin radiates heat energy faster than the shiny tin.

Figure 9.1 Comparing radiation for different surfaces

Absorbing radiation

Dark surfaces are good absorbers of radiation. Very little of the radiation falling on the surface is reflected.

Shiny surfaces are good reflectors of radiation, as they absorb little of the radiation.

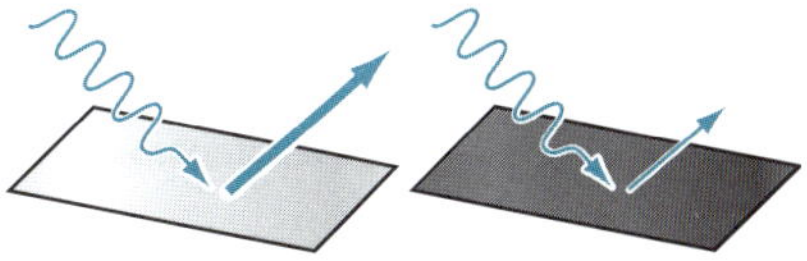

Figure 9.2 Dark surfaces absorb more radiation than shiny surfaces

How buildings lose heat energy

Heat energy from the warm inside of the house is lost to the cool outside by:

- conduction through the solid surfaces, such as walls, window panes, roof tiles, floors
- convection moving the warm air, inside rooms, cavity walls or the roof space, to cooler regions
- radiation from all the outside surfaces of the house.

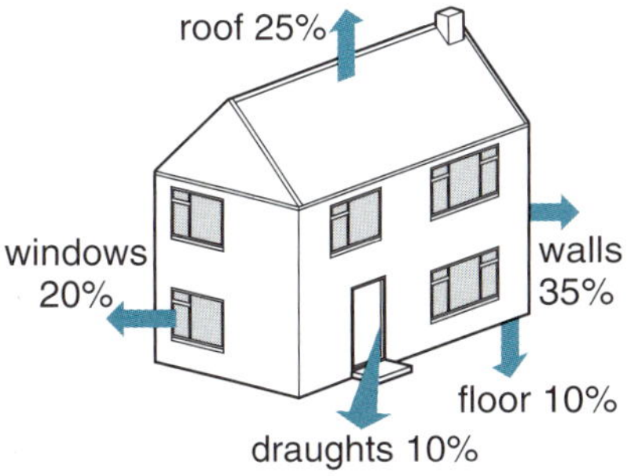

Figure 9.3 Heat loss from a typical house

Reducing the rate of heat loss from buildings

The heat energy is lost less quickly from a warm building by:

- conduction when the solid surface is thicker, or is made from a poor conductor or is in contact with a very bad conductor such as air
- convection when the movement of air is restricted
- radiation when the outside surface is light coloured and shiny.

Insulating buildings

The house in Figure 9.3 will lose less heat energy if it is insulated as shown in the table.

Type of insulation	Cost to install insulation	Savings made to yearly heating bill	Payback time
Double glazing Air between the glass panes cannot move, so reducing heat loss by convection. Also air is a poor conductor/good insulator.	£4000	£200	20 years
Cavity wall insulation Foam fills the space between the two walls. The air pockets in the foam reduce convection and conduction.	£600	£150	4 years
Loft insulation Mineral wool fibres trap air, reducing convection and conduction.	£240	£80	3 years
Floor insulation Thick carpets have air pockets which reduce convection and conduction.	£700	£50	14 years
Draught-proofing Thick curtains stop air moving. The air layer between window and curtain acts as an insulator. Draught excluders stop warm air escaping by convection.	£150	£75	2 years

$$\text{payback time} = \frac{\text{cost of insulation}}{\text{savings made in 1 year}}$$

The most cost effective insulation is the one with the shortest payback time – the draught-proofing. After only 2 years the homeowner has saved on the heating bill the cost of installing the insulation.

CHECK YOUR UNDERSTANDING

❶ Which method of heat transfer does not use particles of matter?

❷ Explain why
a) the bonnets of racing cars are often painted black.
b) houses in hot countries are painted white.

❸ What are the ways of reducing heat loss from the windows of a house?

Transferring electrical energy

Electrical energy is easily transferred as other forms of energy by different devices. Some examples are shown in this table.

Energy input to device	Device	Energy output from device
electrical	heater, immersion heater, cooker, hairdryer	heat (thermal energy)
electrical	filament lamp, TV screen	light energy
electrical	loudspeaker, bell	sound energy
electrical	electric motors, e.g. in a vacuum cleaner, CD driver	movement (kinetic) energy

Many devices are designed to transfer electrical energy as more than one form of energy.

Electrical energy → Hairdryer → Heat (thermal) energy
→ Movement (kinetic) energy

All forms of energy are measured in **joules** (J).

The energy needed to lift an object against the force of gravity is transferred as gravitational potential energy.
For example an electric lift transfers

electric energy ⟶ gravitational potential energy

Gravitational potential energy is the energy an object has due to its height above the Earth's surface.

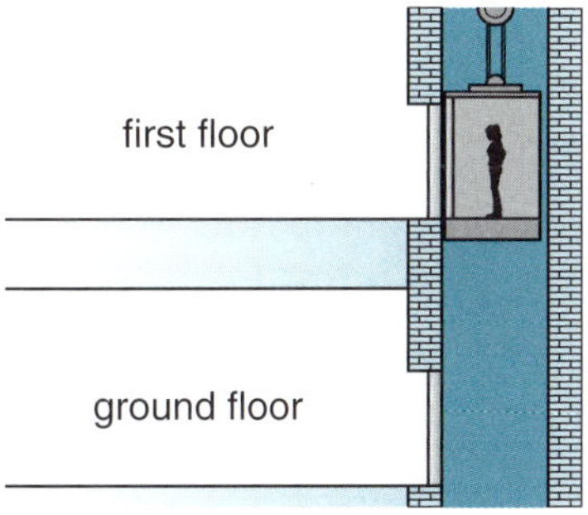

Figure 9.4 A person in a lift on the first floor has gravitational potential energy

How much electrical energy a device transfers depends on:

- the power of the device (how fast the device transfers energy)
- how long the device is used.

$$\text{power} = \frac{\text{energy transferred}}{\text{time taken}}$$

power in watts (W)
energy transferred in joules (J)
time taken in seconds (s)

This equation will be given in an examination question.

1000 W = 1 kilowatt (kW)

1 watt is the transfer of 1 joule of energy in 1 second.

The power of the lift in Figure 9.4 is 8000 W (8 kW). So the lift transfers 8000 J of electrical energy in 1 s.

The amount of energy transferred is calculated using the previous power equation but rearranged.

> energy transferred = power × time taken

This equation will be given in an examination question.

Example: A lift of power 8000 W takes 4 s to move between the first and second floor of a building. How much electrical energy is transferred in this time?

energy transferred = power × time taken
= 8000 × 4
= 32 000 J

Paying for electrical energy

When dealing with large amounts of electrical energy, another unit for energy is used. This is the **kilowatt-hour** (kWh).

> 1 kilowatt-hour is equal to 1000 watts supplied for 1 hour.

Example: Calculate the energy transferred, in kWh, by a 2000 W heater connected to the mains electrical supply for 3 hours.

energy transferred (kilowatt-hours, kWh) = power (kilowatts, kW) × time (hours, h)
= 2 × 3
= 6 kWh

Electricity companies calculate the cost of providing electrical energy using the kilowatt-hour as the unit. If the price for 1 unit is 8p, what is the total cost of using the heater?

> total cost = number of units × cost per unit

This equation will be given in an examination question.

= 6 × 8
= 48p

Example: New meter reading on 1 March: 4 2 6 4 9

Old meter reading on 1 December: 4 0 5 2 6

What is the cost for the units of electricity supplied during these 3 months? The cost of 1 unit = 8p.

number of units = 42649 − 40526 = 2123 units (kWh)
total cost = 2123 × 8 = 16984p = £169.84

CHECK YOUR UNDERSTANDING

1. Name a device that is designed to transfer electrical energy as a) sound energy, b) kinetic energy, c) gravitational potential energy.
2. What is gravitational potential energy?
3. What are the two units used to measure electrical energy?
4. A 500 W drill is used for 5 minutes. How much electrical energy is transferred?

Efficiency

As energy is transferred, only part of it ends up in the form for which the device was designed. This is the **useful energy**. The rest of the energy is **wasted**, usually as heat. The wasted energy spreads out to the surroundings, which become warmer. The wasted energy cannot be saved for further useful energy transfers.

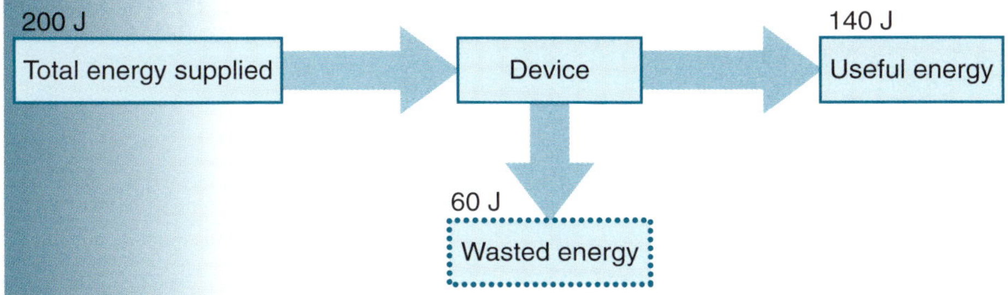

Figure 9.5 Transfer of energy by a device

The more useful energy that the device transfers, the more **efficient** the device.

$$\text{efficiency} = \frac{\text{useful energy transferred by the device}}{\text{total energy supplied to the device}}$$

This equation will be given in an examination question.

Example: An electric motor is supplied with 6000 J of electrical energy. It transfers 4500 J as useful kinetic energy.

i) What is the efficiency of the motor?

ii) How much energy has been wasted?

i) $$\text{Efficiency} = \frac{\text{useful energy transferred by the device}}{\text{total energy supplied to the device}}$$

$$= \frac{4500}{6000}$$

$= 0.75$ (as a percentage, this efficiency $= 0.75 \times 100\%$)

$= 75\%$

ii) The wasted energy = total energy − useful energy = 6000 − 4500 = 1500 J

CHECK YOUR UNDERSTANDING

Lamp	Light energy output in 1 second	Electrical energy input in 1 second
A	0.3 J	15 J
B	1.3 J	20 J

The table compares two lamps.

1. Which lamp is more efficient?
2. What is lamp A's efficiency?
3. How much energy does lamp B waste in 5 seconds?

Environmental effects of using fossil and nuclear fuels

Greenhouse effect

Radiation from the Sun passes through the atmosphere to the Earth's surface. The long wavelength infrared radiation reflected from the surface of the Earth cannot pass through the atmosphere back into space. This energy is absorbed by carbon dioxide (CO_2) in the atmosphere, heating the atmosphere.

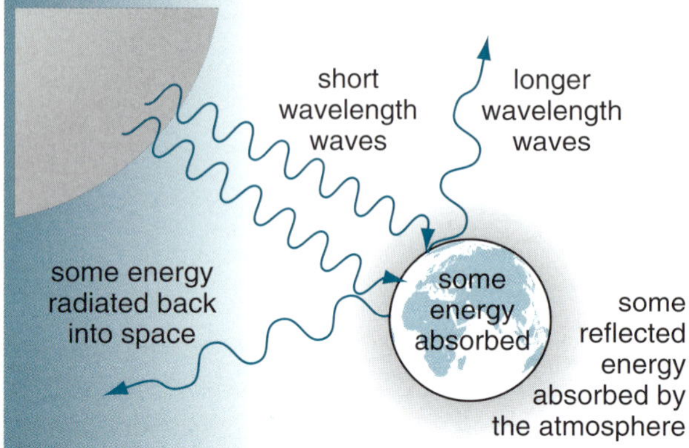

Figure 9.6 Some of the Sun's radiation is reflected back into space

When fossil fuels are burned, the amount of carbon dioxide in the atmosphere increases.

This increases the greenhouse effect, causing increased **global warming**.

Acid rain

When coal and oil are burned, they produce **sulphur dioxide** (SO_2). This gas can be removed before the waste gases from a power station are released into the atmosphere. Or the sulphur can be removed before burning the fuels. However, this increases the cost of producing electricity. Sulphur dioxide dissolves in rain water to form acid rain. Acid rain is harmful to plants and animals.

Nuclear fuels

Nuclear fuels do not produce waste gases that cause acid rain or increase the greenhouse effect. But the waste products from a nuclear power station can be **radioactive** for a very long time. These wastes are stored in strong containers so that they do not contaminate the environment.

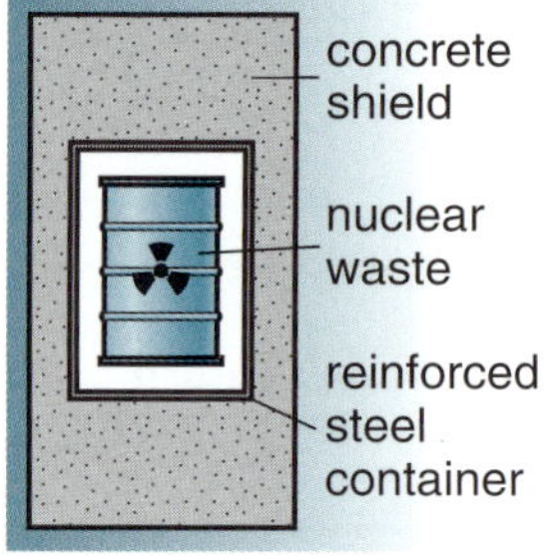

Figure 9.7 Storing nuclear waste

Normally, very little radiation or radioactive material is released by a working nuclear power station or the stored waste. However, in the event of an accident, radioactive material could spread over a very large area and affect it for a long time.

CHECK YOUR UNDERSTANDING

1. Which gases cause:
   a) the greenhouse effect?
   b) acid rain?
2. Why must the waste from nuclear fuels be stored safely in strong containers?
3. Give one reason why sulphur dioxide gas is not removed from the waste gases produced by a coal- or an oil-fired power station.

Generating electricity

Power stations

Most power stations in Britain use the energy from a non-renewable fuel to heat water. Figure 9.8 shows the stages in generating electricity.

Nuclear fuel or fossil fuels:
- coal
- gas
- oil

heat water → **Boiler** producing steam → Steam turns a **turbine** → Turbine turns a **generator**

Figure 9.8

The time it takes to start up a power station so that it begins to generate electricity depends on the type of power station.

shortest start-up time → natural gas, oil, coal, nuclear → longest start-up time

Figure 9.9

A **geothermal** power station uses the hot water in the rocks from certain areas just under the Earth's surface. This water rises to the surface as steam, and is used directly to drive turbines. Sometimes this renewable resource is only used to provide local heating.

Other **renewable energy resources** provide movement (kinetic energy) to turn the turbines.

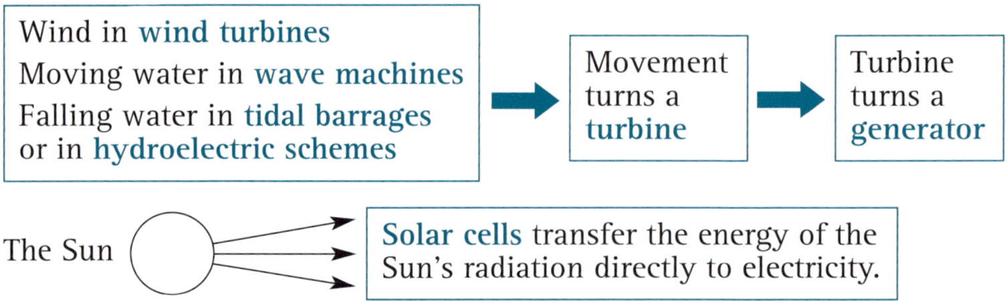

Figure 9.10 How renewable resources generate electricity

The table below compares the advantages and disadvantages for each resource used in the large-scale production of electric power.

Resource	Advantages	Disadvantages
Nuclear fuels	No polluting gases produced. A small amount of fuel gives a large amount of energy.	Non-renewable energy resource. Some radioactive waste must be stored for thousands of years. Serious accidents may release radiation over a large area.
Fossil fuels – coal, oil, natural gas	Produce electricity when it is needed, they do not depend on the weather. Gas-burning power stations can be switched on and off quickly.	Non-renewable energy resource. Burning coal, oil and natural gas adds carbon dioxide, a greenhouse gas, to the atmosphere. Burning coal and oil adds sulphur dioxide to the atmosphere causing acid rain.
Wind	Renewable energy resource. Low running costs. No polluting gases produced.	Wind generators can spoil the landscape and cause unwanted noise. Unreliable – electricity is only generated when there is a strong enough wind.
Tides	Renewable energy resource. Low running costs. No polluting gases produced.	Very expensive to build. Destroys the habitat of wading birds. Electricity is only generated at certain times of the day.
Hydroelectric	Renewable energy resource. Can be switched on and off quickly. Can be used to store energy. No polluting gases produced.	Large areas of land may be flooded.
Waves	Renewable energy resource. Low running costs. No polluting gases produced.	Difficult to build a power station strong enough to stand up to the power of rough seas.
Solar	Renewable energy resource. Ideal for remote places. Ideal when only a small amount of electricity is needed. No polluting gases produced.	Expensive to set up in large numbers. The amount of electricity produced depends on the brightness of the light.
Geothermal	Massive amounts of energy are available.	Unless there are natural geysers it can be expensive and inefficient to generate electricity.

CHECK YOUR UNDERSTANDING

❶ Which energy resource:
a) reduces land space for forestry or farming?
b) depends on a monthly and yearly cycle?
c) allows a fossil fuel power station to start up most quickly?

❷ Which power stations are used to supply electricity at peak demands?

❸ Give an example of where a solar panel is used even though it is at high cost.

Topic Test

1 Match each electrical device with its useful energy transfer.

Electrical device	Useful energy transfer
[A] food mixer	heat energy [E]
[B] computer screen	light energy [F]
[C] loudspeaker	movement energy [G]
[D] storage radiator	sound energy [H]

(4 marks)

2 The diagram shows the stages in a coal-fired power station. Match these names with the correct label box:

boiler coal generator steam turbine

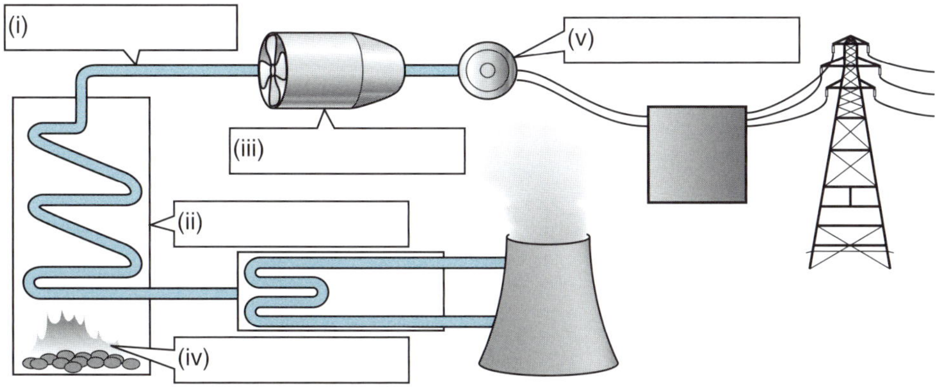

(5 marks)

3 a) Copy and complete these sentences explaining why an attic room feels cold in the winter and hot in the summer.

In winter, the heat energy is i)______ from the room to the ii)______ outside mainly by iii)______ and iv)______ from the roof. The rate of energy entering the attic is v)______ than the rate of energy transfer out of the attic, so the temperature vi)______. In summer, the roof vii)______ energy from the Sun at a viii)______ rate than energy going out, so the attic becomes ix)______.

(9 marks)

b) The house has an insulating lining fitted to the inside of the roof. This cost £300. In one year, there is a saving of £60 to the heating bill for the attic room. How long will it take to pay back the initial cost? (2 marks)

4 A housing estate is provided with hot water and heating from a central supply. This is attached to an oil-fired power station.

a) How does this increase the efficiency of the power station?

b) What is the environmental advantage in using such a scheme?

c) What problems will there be in transferring hot water to the houses?

d) Describe how these problems can be solved. (5 marks)

5 a) What is the primary energy source for a hydro-electric scheme?

One such scheme generates 200 kW of electric power from a total available power of 500 kW.

b) What is the electrical energy transferred in 1 second?

c) What is the energy wasted in 1 second?

What is the efficiency of this scheme?

$$\text{energy transferred} = \text{power} \times \text{time}$$

$$\text{efficiency} = \frac{\text{useful energy transferred by device}}{\text{total energy supplied to device}}$$

(4 marks)

6 The cost of using electricity is 7p for 1 unit. From these readings find a) the amount of units used, b) the total cost.

$$\text{total cost} = \text{number of units} \times \text{cost per unit}$$

New meter reading	Old meter reading
6 8 9 5 7 kWh	6 7 4 3 2 kWh

(3 marks)

Total: 32 marks

ELECTRICITY

Circuits

The table below summarises some information you should already know.

Quantity	Unit	Measured using
potential difference (V) or voltage	volt (V)	a voltmeter joined across the component
current (I)	ampere (A) or amp	an ammeter joined in series with the component

REMEMBER – Potential difference (p.d.) is the same as voltage.

LEARN these circuit symbols. You may be asked to draw and/or interpret circuit diagrams using them.

Circuit symbols

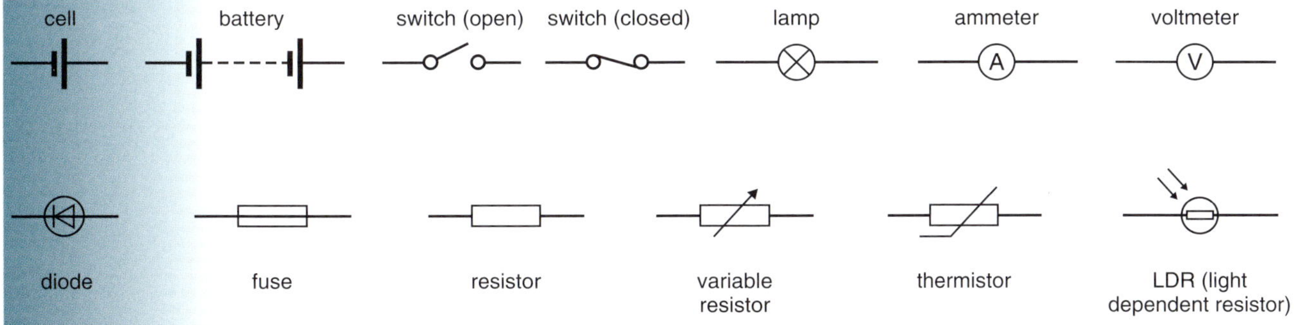

Figure 10.1 Electrical components and their symbols

REMEMBER – A battery is two or more cells joined together.

Series circuits

The current in a **series circuit** has only one path it can take. So the same current (I) must flow through each component and wire in the circuit.

Adding more components to the circuit will increase the resistance of the circuit and decrease the current.

The total resistance of a series circuit is worked out by adding together the resistances of all the individual components.

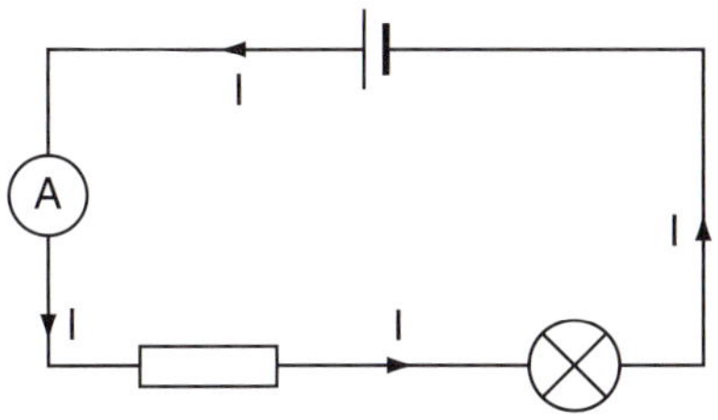

Figure 10.2 A series circuit

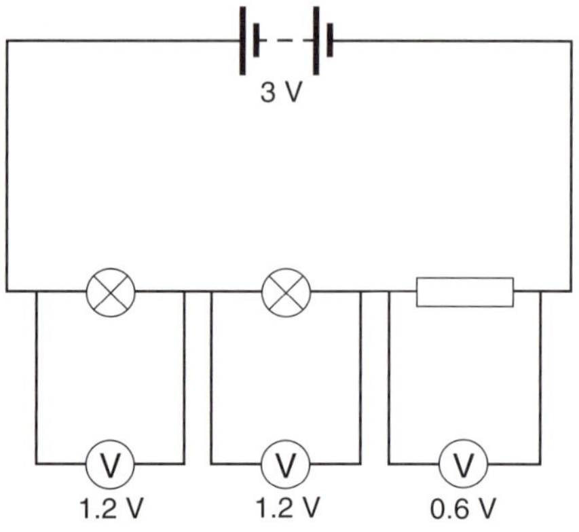

Figure 10.3 Measuring the p.d. across components in a series circuit

When components are joined in series, the potential difference (p.d.) of the power supply is shared between the components in the circuit.

> total p.d. across all components = p.d. of the power supply

When cells are joined in series, the total potential difference is the sum of the individual potential differences. Remember, this only works if the cells are joined correctly, positive (+) to negative (−).

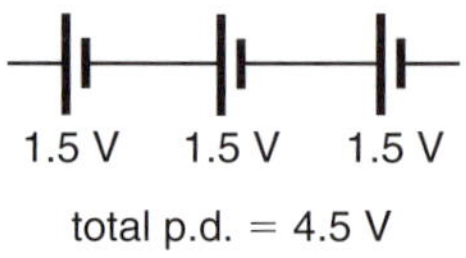

Figure 10.4 Cells joined correctly

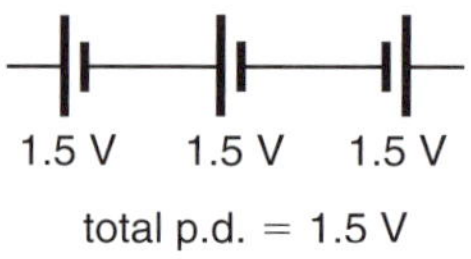

Figure 10.5 One cell is the wrong way round

Parallel circuits

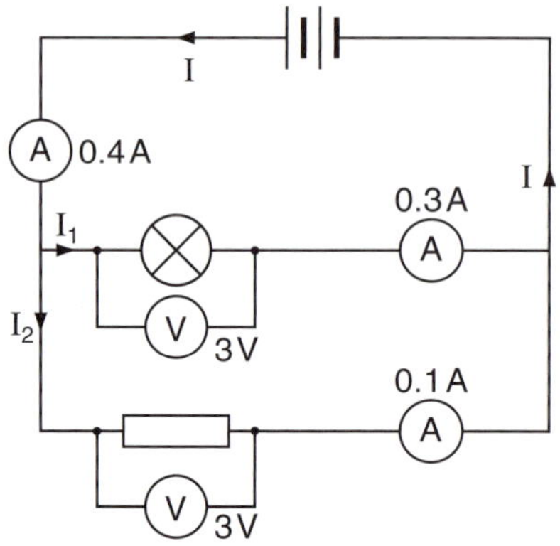

Figure 10.6 A parallel circuit

For components joined in parallel:

- The potential difference across each component is the same.
- The bigger the resistance of a component, the smaller the current flowing through it.
- The total current flowing in the whole circuit is the sum of the currents through the individual components.

  e.g. $I = I_1 + I_2$
  $0.4\text{ A} = 0.3\text{ A} + 0.1\text{ A}$

Resistance

The current through a component depends on the potential difference across the component and the resistance of the component.

potential difference = current × resistance
potential difference in volts (V)
current in amperes (A)
resistance in ohms (Ω)

Resistance is measured in **ohms** (Ω).

LEARN THIS EQUATION and the UNITS.

Light dependent resistor (LDR)

The resistance of a light dependent resistor goes down as the intensity (brightness) of the light hitting it goes up.

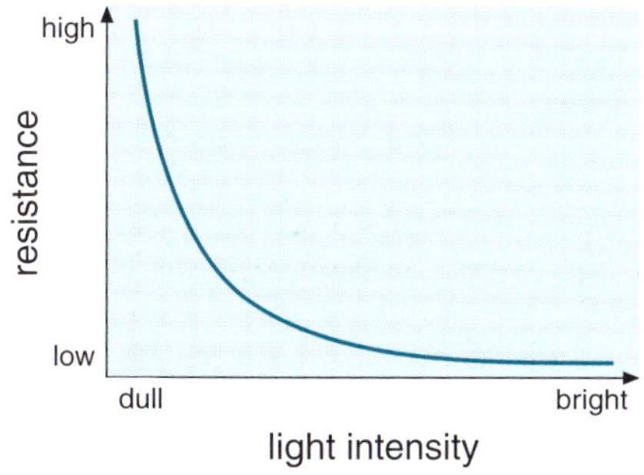

Figure 10.7 Variation of resistance with light intensity for an LDR

Thermistor

The resistance of a thermistor goes down as the temperature of the thermistor goes up.

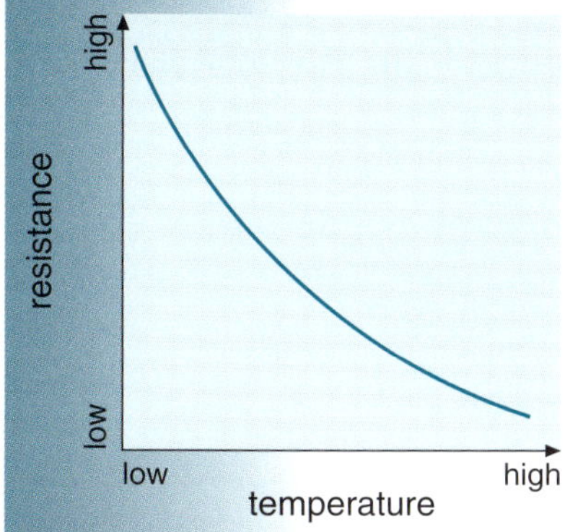

Figure 10.8 Variation of resistance with temperature for a thermistor

Current–voltage graphs

The circuit in Figure 10.9 can be used to investigate how the current through a component depends on the voltage (p.d.) across the component.

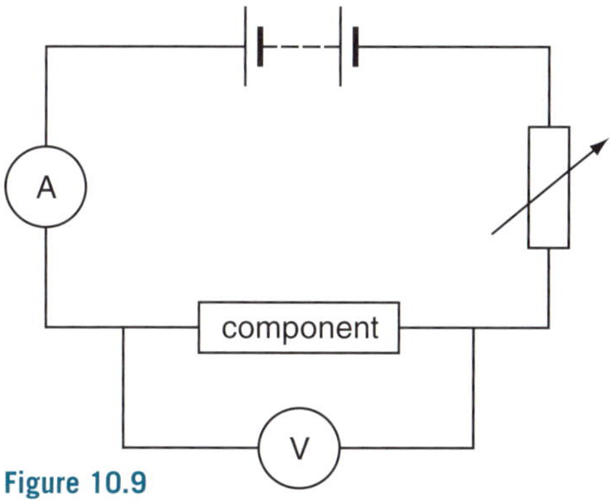

Figure 10.9

The circuit can also be used to measure the resistance of a component.

Resistor at constant temperature

The current through a resistor is directly proportional to the voltage across the resistor. (So the resistance is constant.)

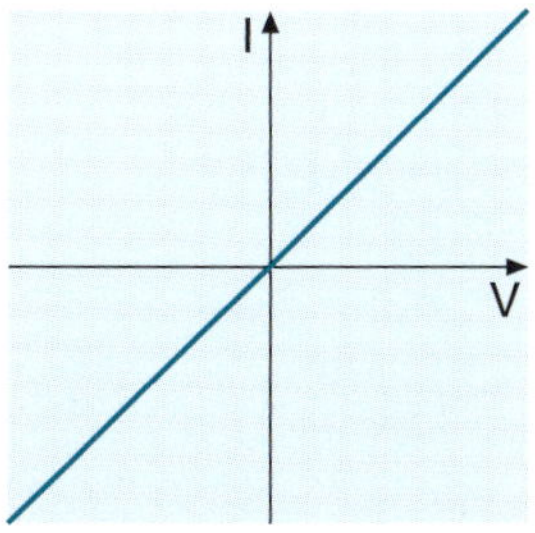

Figure 10.10 Current and voltage for a resistor

Filament lamp

Increasing the voltage makes the filament in the lamp brighter and hotter. As the temperature of the filament goes up, the resistance of the lamp goes up.

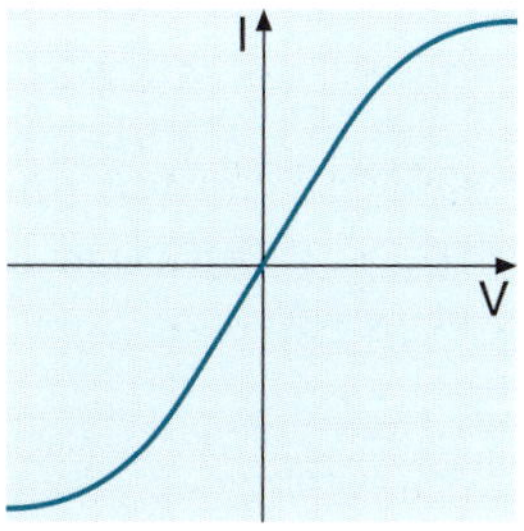

Figure 10.11 Current and voltage for a filament lamp

Diode

Diodes only let current flow in one direction. In the reverse direction diodes have a very high resistance so no current flows.

You should be able to recognise and/or draw the graphs in Figures 10.10–10.12.

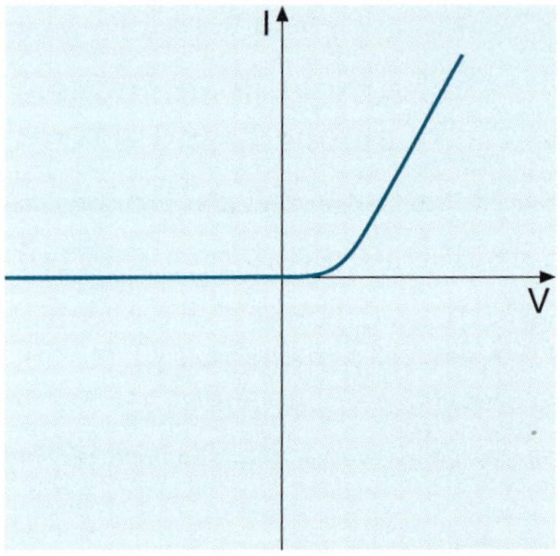

Figure 10.12 Current and voltage for a diode

CHECK YOUR UNDERSTANDING

1. Why can a diode be used to stop current flowing the wrong way in a circuit?
2. The current through a thermistor at 40°C is 0.02 A. The resistance of the thermistor is 600 Ω.
   a) Calculate the voltage across the thermistor.
   b) What happens to the resistance of the thermistor as its temperature goes up?
3. Draw (without looking) the current-voltage graph for a filament lamp.
4. How does the current through an LDR change with light intensity?

Electricity and magnetism

When an electric current flows through a coil of wire, the coil becomes an electromagnet. Reversing the direction of the current reverses the north and south poles of the electromagnet.

The d.c. electric motor

When a current flows through the coil it becomes an electromagnet. This magnetic field interacts with the permanent magnetic field. Each side of the coil experiences a force, but in opposite directions. So one side goes up, the other side goes down and the coil turns. This is the basis of an electric motor.

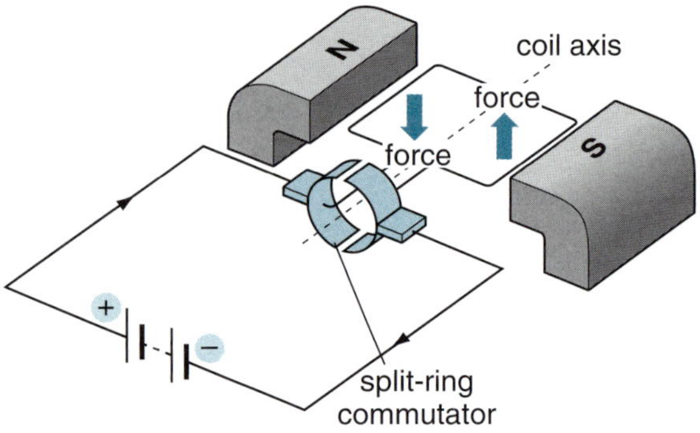

Figure 10.13 A d.c. electric motor

The circuit breaker

This is a switch that breaks a circuit when the current gets too high. The type of circuit breaker in Figure 10.14 uses an electromagnet. If a fault makes the current increase, the electromagnet becomes strong enough to pull across the iron catch. This opens the contacts, switching off the circuit. Once the fault is repaired, the contacts are closed by pressing the reset button.

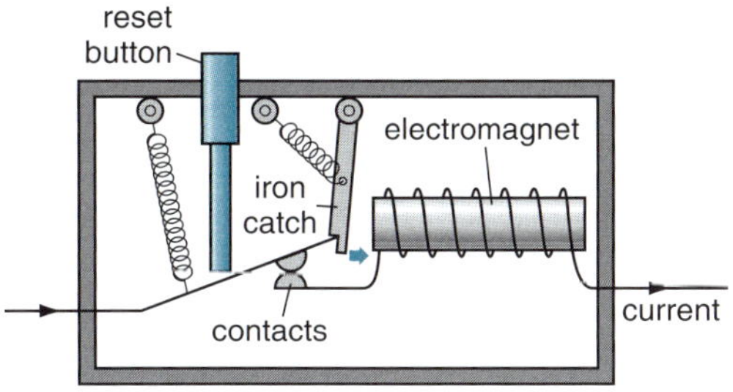

Figure 10.14 A circuit breaker

CHECK YOUR UNDERSTANDING

1. Explain why the coil in an electric motor turns when a current flows through it.
2. Suggest why it may be better to use a circuit breaker rather than a fuse in a mains circuit.

Static electricity

Rubbing different insulating materials together will usually make them become electrically charged. This is because electrons move off one material on to the other. The material that loses electrons becomes positively charged and the one gaining electrons negatively charged.

Charging a polythene rod

The rod gains electrons from the cloth. The rod becomes negative.

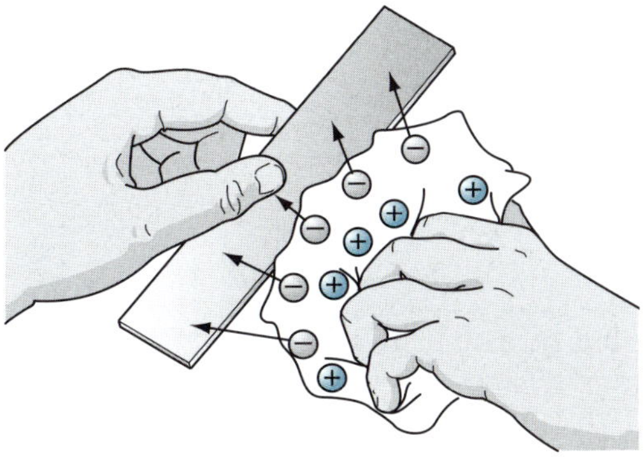

The cloth loses electrons. The cloth becomes positive.

Figure 10.15 Charging a polythene rod

Charging a Perspex rod

The rod loses electrons to the cloth. The rod becomes positive.

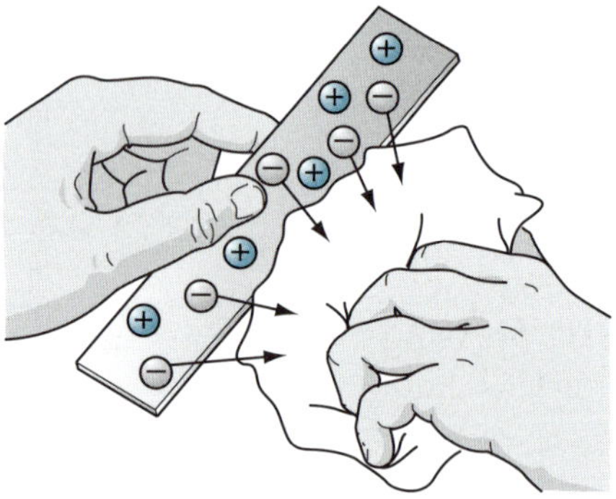

The cloth gains electrons. The cloth becomes negative.

Figure 10.16 Charging a Perspex rod

Charged objects exert forces on each other:

- Objects with the **same charge repel** (+ and + OR − and −).
- Objects with **opposite charge attract** (+ and −).

If the objects are close enough, the forces may make them move. Charged objects can also attract small uncharged objects that are nearby. For example, a charged nylon comb will attract dry hair.

Uses of static electricity

Electrostatic smoke precipitator

This is used to remove smoke from waste gases.

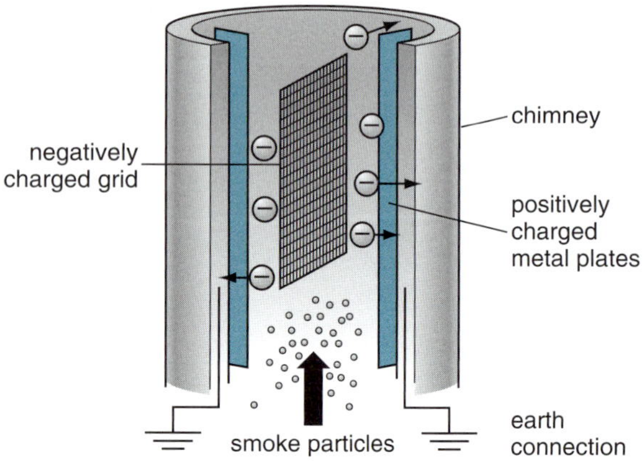

Figure 10.17 An electrostatic smoke precipitator

1 The gases and smoke pass a charged grid.

2 The smoke particles gain the same charge as the grid.

3 The smoke particles are repelled by the grid and attracted to the oppositely charged metal plates where they stick to the plates.

4 The metal plates are knocked so the smoke particles fall and can be collected.

Making a photocopy

1 The light-sensitive (photoconductive) copying plate is given a charge.
2 The page to be copied is placed above the plate and lit with a strong light.
3 Where light hits the plate, the charge leaks away.
4 The pattern of the charge left on the plate is the same as the dark parts of the original page.
5 Black powder (toner) is attracted to the charge still on the plate.
6 The toner is transferred to a sheet of paper.
7 The paper is heated to make the toner stick.
8 The sheet of paper is a photocopy of the original.

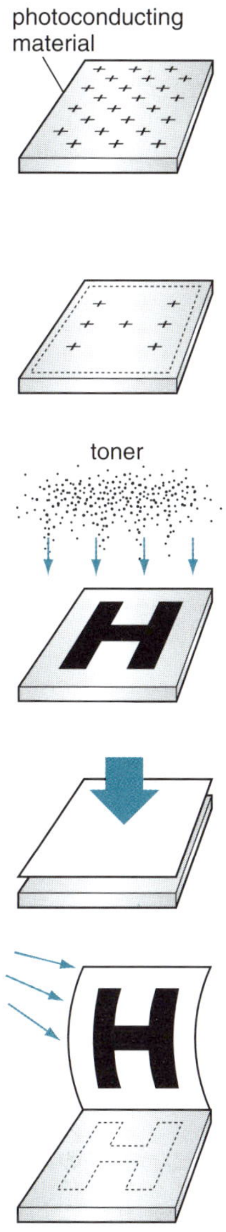

Figure 10.18 Making a photocopy

Dangerous situations

If static electricity is allowed to build up, the charge could cause a spark. The spark could then cause an explosion.

Delivering fuel

Tankers deliver large amounts of fuel to garages and aircraft. As the fuel flows through the delivery pipe both the pipe and the fuel become charged. To stop the charge building up, the tanker is **earthed** (joined to the ground by a metal wire). The charge flows safely through the wire to earth.

Lightning

A tall building is usually protected by a **lightning conductor**. This is a thick strip of copper (a good conductor) that runs from the roof of the building into the ground. If lightning did strike, the charge would flow through the lightning conductor to earth without damaging the building.

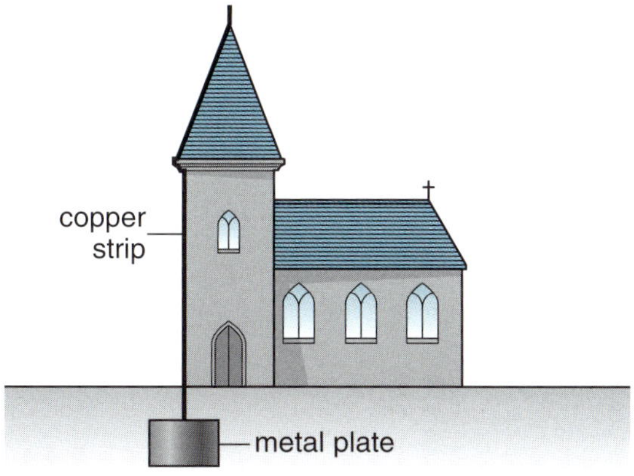

Figure 10.19 A lightning conductor

CHECK YOUR UNDERSTANDING

❶ Explain what is meant by something being 'earthed'.

❷ Will two charged ebonite rods placed close together attract or repel? Explain the reason for your answer.

❸ A charged balloon is placed close to some small pieces of paper. What happens?

Mains electricity

Many electrical appliances work by plugging them into the 230 volt mains electricity supply. The cable joining the appliance to the plug and the plug itself are designed to make the appliance safe to use.

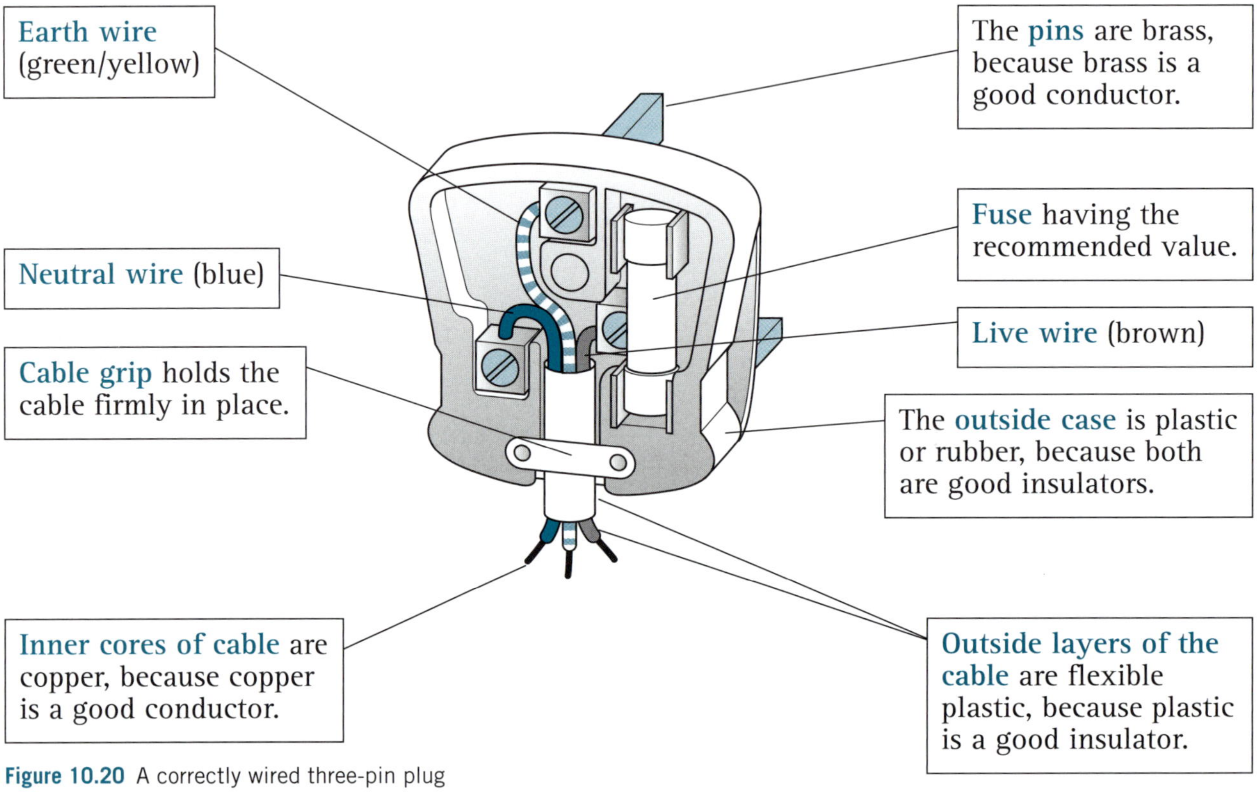

Figure 10.20 A correctly wired three-pin plug

You should be able to spot any mistakes that have been made in wiring a plug. For example, the plug on the right has no cable grip, also the live and neutral wires are connected to the wrong terminals.

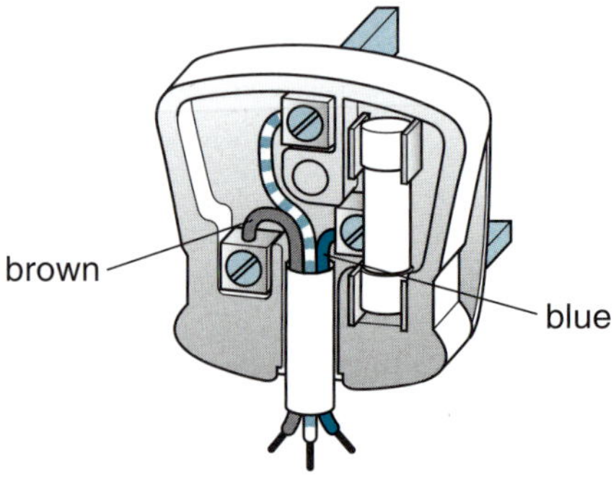

Figure 10.21 An incorrectly wired plug

Fuses

A fuse is a thin piece of wire that lets a current up to a certain value flow through it. Above this value the fuse will overheat and melt. This breaks the circuit and switches off the current. To protect an appliance, the fuse should have a value close to but higher than the normal working current of the appliance.

Earthing

All appliances with a metal case should be earthed. The earth wire (yellow/green) joins the earth pin of the plug to the metal case of the appliance.

The appliance is switched on. → A fault connects the live wire to the metal case. → A large current flows to earth. → The fuse melts breaking the circuit.

Mains electricity is dangerous; it is important it is used correctly. You should be able to recognise from diagrams situations that could be dangerous. For example, switching an appliance on or off with wet hands.

Alternating current (a.c.)

An **alternating current** (a.c.) in a circuit is constantly reversing direction. The number of times the current (or voltage) changes direction in one second gives the frequency of the supply.

> Frequency of the UK mains supply = 50 hertz (Hz)

An **oscilloscope** (CRO) is used to compare the frequency and peak voltage (maximum) of a.c. supplies. (The controls on the CRO must stay the same.)

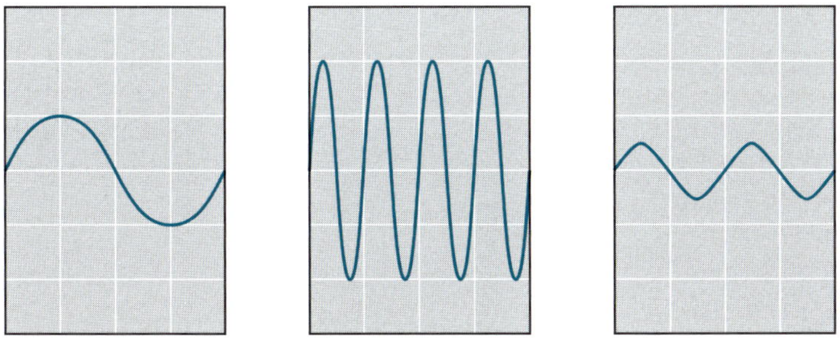

Figure 10.22 CRO traces

Direct current (d.c.)

A **direct current** (d.c.) always flows in the same direction. A cell or battery gives a steady direct current.

The oscilloscope (CRO) trace for a steady d.c. supply is a horizontal line. The bigger the voltage of the supply, the further the line is from the centre of the screen.

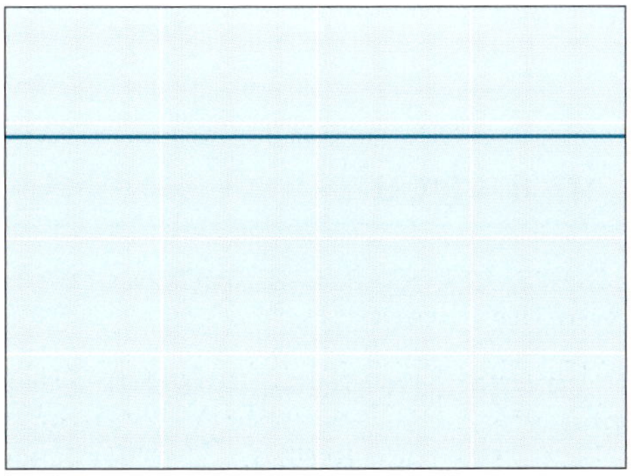

Figure 10.23 The wave trace on a CRO for a d.c. supply

CHECK YOUR UNDERSTANDING

1. What is the colour of the insulation on the neutral wire inside a plug?
2. Why is brass used to make the pins of a plug?
3. What material is used to make the case of a plug?
4. a) Draw the oscilloscope trace for an a.c. supply.
   b) How would the trace change if the frequency of the supply was doubled?
5. Explain how a fuse works.

Electrolysis

> An **ion** is an atom that has lost or gained one or more electrons.

Some chemical compounds conduct electricity when they are melted or dissolved in water. The compounds conduct because they contain ions. The negatively charged ions move to the positive electrode, and the positively charged ions move to the negative electrode. When the ions reach the electrodes, simpler substances, either gases or solids, are released.

During the electrolysis of copper sulphate solution the substances released are:

- copper (solid) at the negative electrode.
- (gas) at the positive electrode.

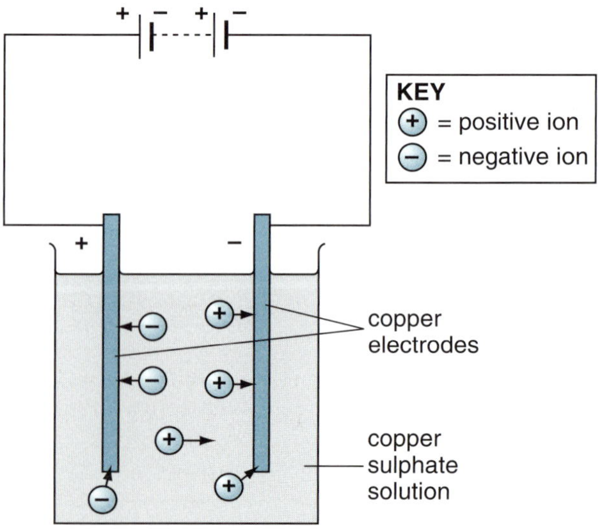

Figure 10.24 Electrolysis of copper sulphate

Energy and power in a circuit

> An electric current is a flow of charge (electrons or ions).

> 1 watt of power is the transfer of 1 joule of energy in 1 second.

When an electric current flows, energy is transferred from the power supply to the components in the circuit.

When the component that charge flows through is a resistor, electrical energy is transferred as heat.

The **power** of an appliance or component is the **rate at which it transfers energy**.

power = potential difference × current
power in watts (W)
p.d. in volts (V)
current in amperes (A)

LEARN THIS EQUATION and the UNITS

Induced current

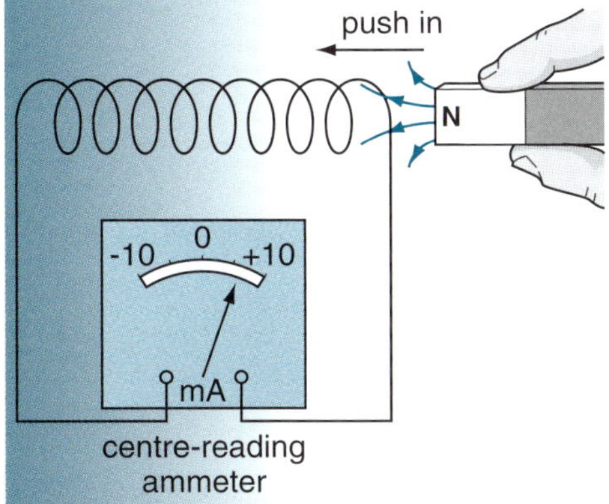

Figure 10.25

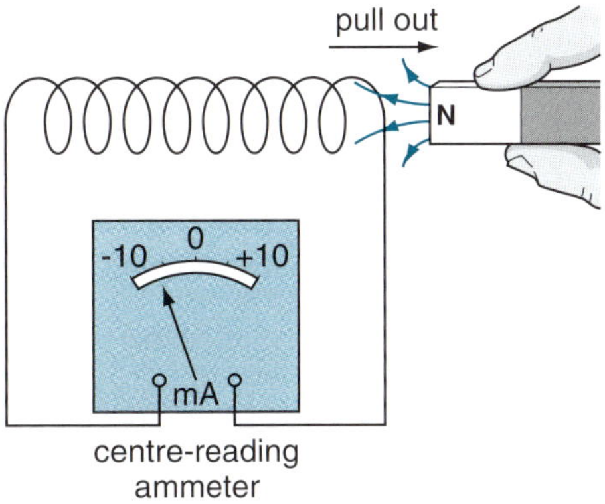

Figure 10.26

Pushing a magnet into a coil of wire will produce (induce) a current in the wire. For this to happen the coil must be part of a complete circuit.

Pulling the magnet out of the coil or pushing the other pole of the magnet into the coil reverses the direction of the induced current.

No current is induced when the magnet is not moving.

Generating electricity

Electricity is generated using the induction effect. This can be done in two ways.

- A coil of wire is rotated in a magnetic field.
- A magnet is rotated inside a coil of wire.

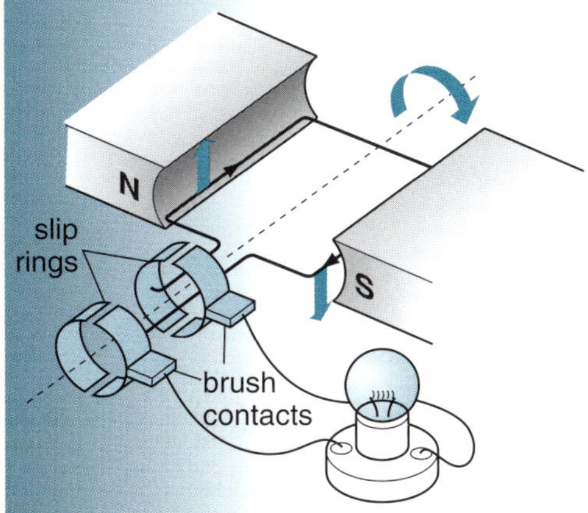

Figure 10.27

The coil rotating between the north and south poles of the magnet cuts through magnetic field lines. This induces a voltage in the coil which, if the coil is part of a complete circuit, causes a current to flow.

The size of the induced voltage (and current) can be increased by:

- rotating the coil faster
- using a stronger magnet
- having more turns on the coil
- increasing the area of the coil.

Transformers

A **transformer** is used to increase or decrease the voltage of an **a.c. supply**. Transformers are an essential part of the electricity distribution system. The network of power lines is called the **National Grid**.

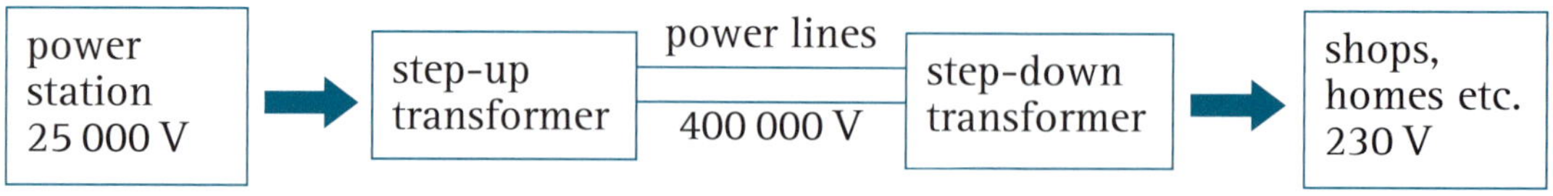

CHECK YOUR UNDERSTANDING

1. A radio draws a current of 0.4 A from a 9 V battery. Calculate the power of the radio.
2. A current is induced in a coil of wire as it rotates between the poles of a magnet. What changes can be made to increase the induced current?
3. What happens at the two electrodes during electrolysis?
4. What energy transfer occurs when a current flows through a resistor?

Topic Test

1 Match each symbol to the correct component listed in the box below. Each component should be used once or not at all.

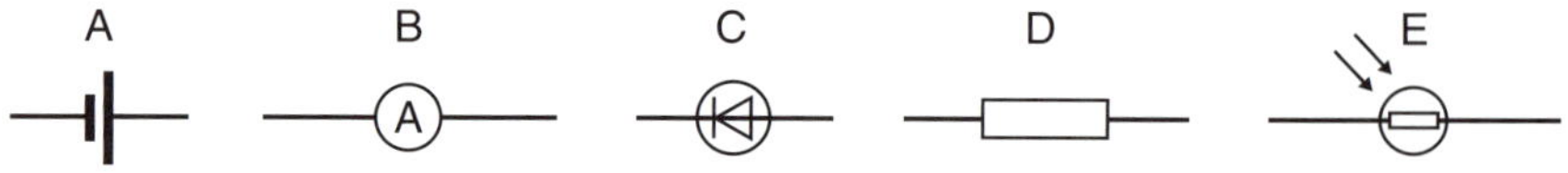

ammeter	battery	cell	diode	LDR	resistor	thermistor

(5 marks)

2 Copy and complete the following sentences by choosing the correct words from the box. Each word should be used once or not at all.

conductor	copper	earth	fuse	insulator	live	neutral	plastic

In a 3-pin plug the brown wire is joined to the ______________ terminal through the ___________. The blue wire is attached to the ___________ terminal. The case of the plug is made of ____________ because it is a good _______________. The core of the cable is made of _____________.

(6 marks)

3 This question is about a hairdryer. Match each quantity in the box to the correct space in the sentences.

6 amps	50 hertz	230 volts	1380 watts

The hairdryer plugs into the mains supply of ____________. The frequency of the mains supply is ____________. The hairdryer has a power of ____________ and draws a current of ____________ from the supply. (4 marks)

4 For safety, an electric iron is fitted with an earth wire. Explain how the earth wire acts as a safety device. (3 marks)

5 An electric fire draws a current of 8 amps from the 230 volt power supply.

a) Calculate the power rating of the fire.

b) Calculate the resistance of the fire's heating element when it is working normally.

(6 marks)

6 Explain how a smoke precipitator inside a power station chimney separates the soot and ash particles from waste gases. (3 marks)

7 A coil of wire is connected to a sensitive ammeter. When the south pole of a bar magnet is pushed slowly into the coil, the needle on the ammeter deflects two scale divisions to the left. What would happen if:

a) The north pole of the magnet was pushed at the same speed into the coil?

b) The south pole of the magnet was pulled out of the coil very fast?

c) The magnet was held still inside the coil? (3 marks)

Total: 30 marks

FORCES

Measuring and representing motion

Velocity and acceleration

Speed is the distance an object travels in a given time.

Speed is measured in metre/second (m/s).

$$\text{speed} = \frac{\text{distance}}{\text{time}}$$

Velocity

Speed and velocity are *not* the same thing. To give velocity you must give both speed and direction. This is because the velocity of an object is its **speed in a given direction**.

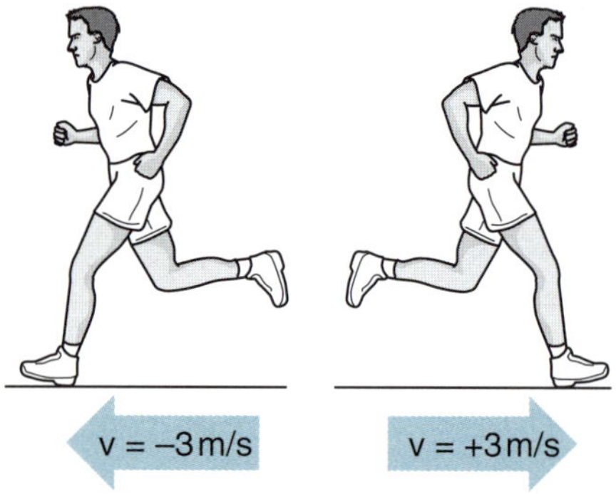

The two joggers are running at the same speed but in different directions. So they have different velocities.

Figure 11.1 These joggers have different velocities

Acceleration

An object is accelerating when its velocity is changing. The faster the velocity changes, the larger the acceleration.

$$\textbf{acceleration} = \frac{\textbf{change in velocity}}{\textbf{time taken}}$$

velocity in metres/second (m/s)
time taken in seconds (s)
acceleration in metres/second squared (m/s^2)

LEARN THIS EQUATION and the UNITS.

When objects move in a straight line, 'change in velocity' and 'change in speed' are the same thing.

Example: At the start of a race a cyclist accelerates to 15 m/s in 3 s. What is the acceleration of the cyclist?

$$\text{acceleration} = \frac{\text{change in velocity}}{\text{time taken}} = \frac{15 - 0}{3} = 5\ \text{m/s}^2$$

Always start a calculation with an equation and finish with a unit.

Distance–time graphs

A distance–time graph can be used to describe the speed of an object.

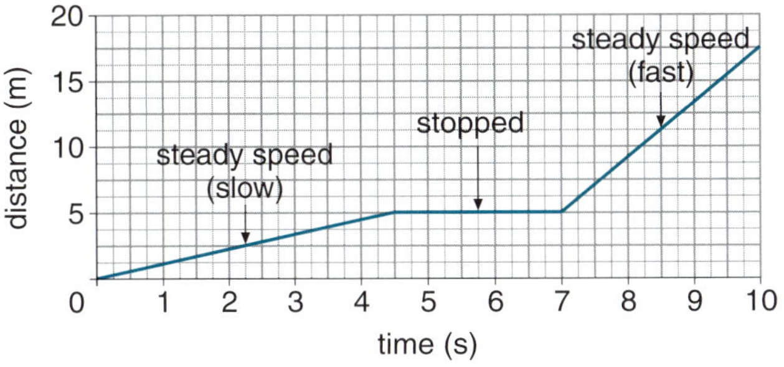

A flat (horizontal) line means the object is not moving.

A straight line going up means the object is moving with a steady speed.

The steeper the slope (gradient) of the line, the faster the speed of the object.

Figure 11.2 A distance–time graph

Velocity–time graphs

A velocity–time graph can be used to describe velocity and acceleration.

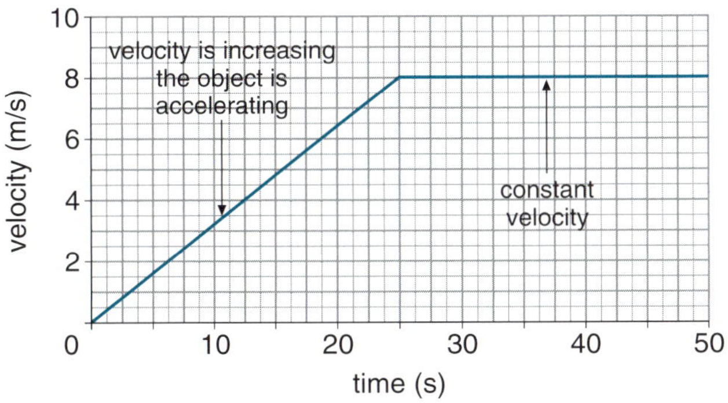

A flat (horizontal) line means the object is moving with a constant velocity.

Remember – Constant velocity means both speed and direction are not changing.

A straight line going up means the object is accelerating.

The steeper the slope (gradient) of the line, the larger the acceleration of the object.

Figure 11.3 A velocity–time graph

CHECK YOUR UNDERSTANDING

1. How is velocity different from speed?
2. Draw a distance–time graph for a car moving at constant speed along a straight road.
3. The velocity–time graph shown is for a motorcyclist travelling along a straight road. Describe the motion of the motorcyclist.

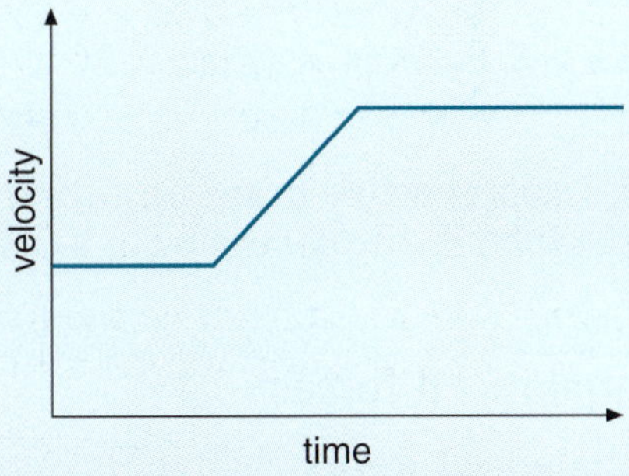

Balanced and unbalanced forces

All forces have a size, measured in newtons (N), and a direction. When more than one force acts on an object, the forces can balance and cancel out.

Balanced forces

Balanced forces do not change the movement of an object.

If an object is stationary (not moving), it will remain stationary. So if the bag of shopping in the picture is not moving, the forces must be balanced.

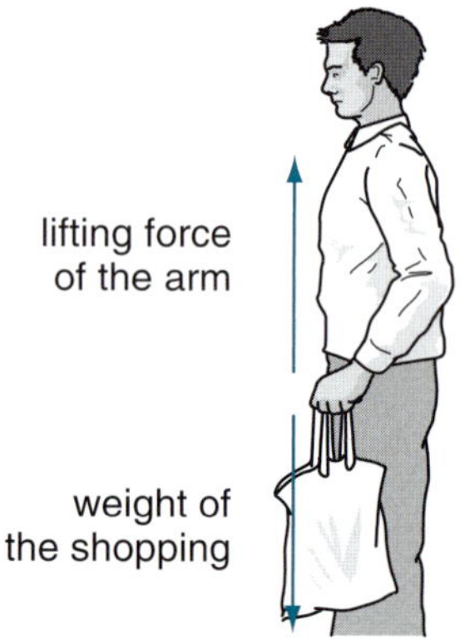

Figure 11.4

> lifting force upwards = weight downwards

When the forces on a moving object are balanced, the object will keep moving at the same speed in the same direction.

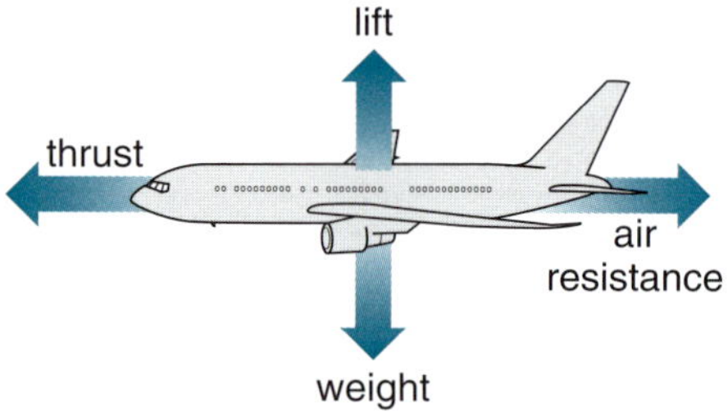

Figure 11.5 Forces on an aircraft

Figure 11.5 shows the forces on a flying aircraft. When the force down balances the force up and the force forward balances the force backward, the aircraft will be flying at a constant height with a constant speed.

Air resistance (or drag), is a force of friction. The direction of air resistance is always opposite to the direction of the moving object.

Unbalanced forces

Unbalanced forces change the movement of an object. An unbalanced force can make an object speed up or slow down.

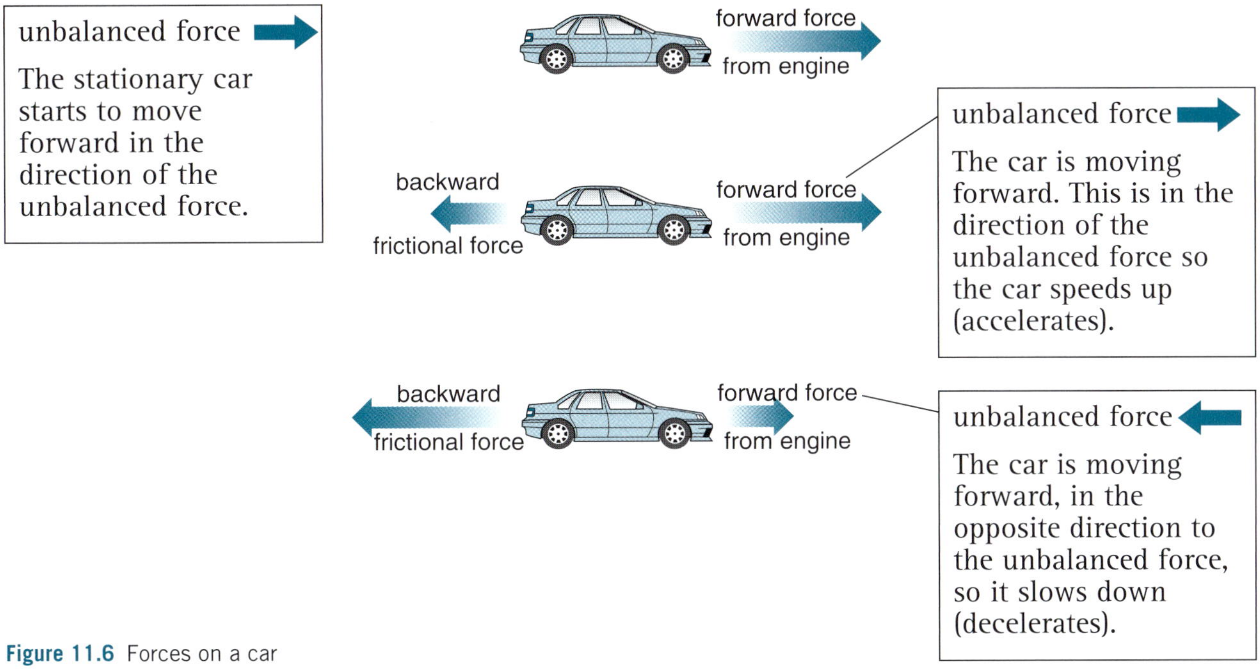

Figure 11.6 Forces on a car

An object will only accelerate when an unbalanced force acts on it. It then accelerates in the direction of the unbalanced force.

The bigger the force on the object, the bigger the acceleration of the object.

The bigger the mass of an object, the bigger the force needed to make it accelerate.

Falling bodies

Gravity will cause a falling body to accelerate. **Weight** is the force due to gravity.

On Earth, gravity pulls on every one kilogram of mass with a force of about 10 newtons. This is called the **gravitational field strength** (g).

$$g = 10 \text{ N/kg}$$

weight = mass × gravitational field strength
weight in newtons (N)
mass in kilograms (kg)
gravitational field strength in newtons/kilogram (N/kg)

LEARN THIS EQUATION and the UNITS.

Forces on a skydiver

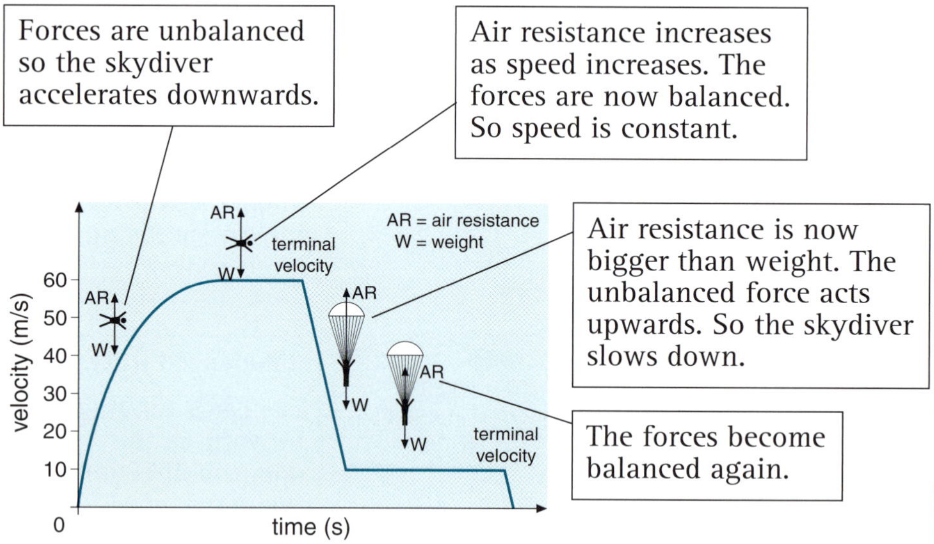

Figure 11.7 Graph showing the forces on a skydiver

The constant speed reached by a falling body is called its **terminal velocity**.

CHECK YOUR UNDERSTANDING

1. A car has a mass of 800 kg. What does it weigh?
2. Explain why a ball thrown up into the air slows down.
3. Explain why a skydiver reaches a terminal velocity before the parachute is opened.
4. Explain why a lorry needs more powerful brakes than a car.

Stopping safely

The total stopping distance of a vehicle is made up of two parts: **thinking distance** and **braking distance**.

stopping distance = thinking distance + braking distance

Thinking distance is how far a vehicle travels during the time it takes the driver to react to a hazard and apply the brakes.

Braking distance is how far a vehicle travels before stopping, after the brakes have been applied.

Speed changes the stopping distance. The faster the vehicle, the longer the stopping distance. Figure 11.8 shows the average stopping distance for a family car at different speeds. The braking force is the same for each speed.

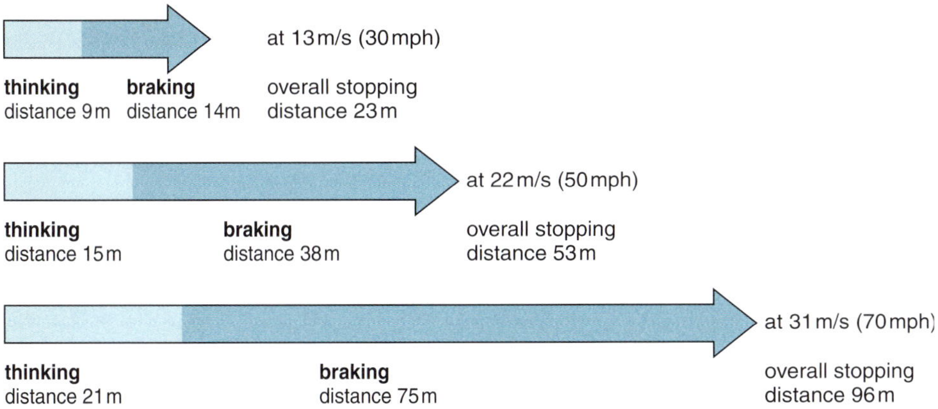

Figure 11.8 Stopping distances at different speeds

The stopping distance is also affected by:

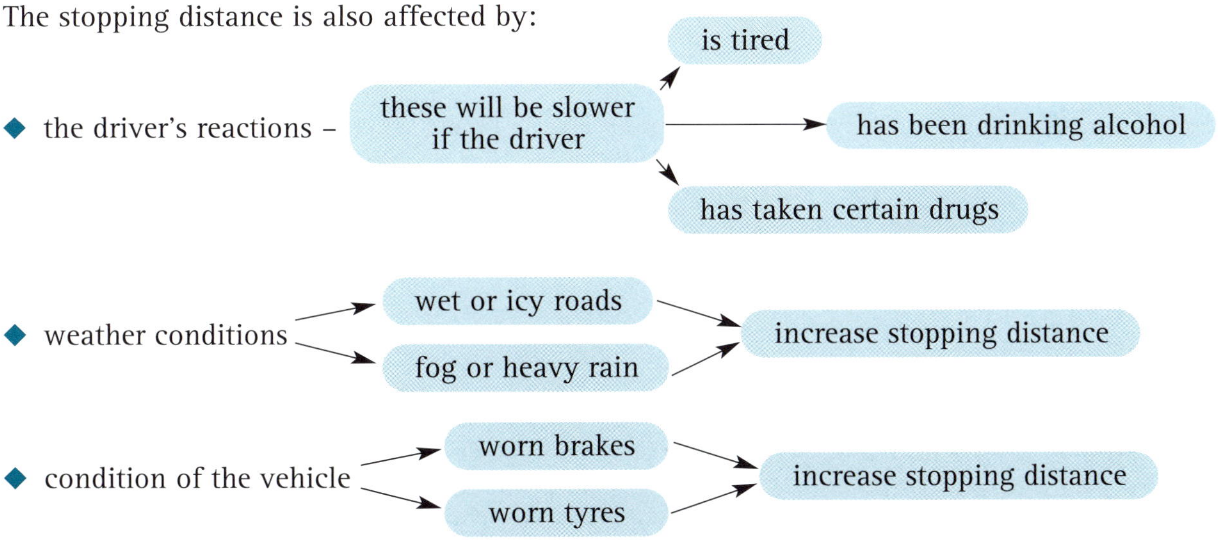

Work and energy

Forces that make something move are doing **work**. The amount of work done depends on:

- the size of the force
- the distance the force moves.

The bigger the force, and the further it moves, the more work is done.

work done = force × distance moved (in the direction of the force)
work done in joules (J)
force in newtons (N)
distance moved in metres (m)

LEARN THIS EQUATION and the UNITS.

Example: How much work is done by a high jumper lifting her own weight of 480 N a height of 1.2 m?

$$\text{work done} = \text{force} \times \text{distance moved}$$
$$= 480 \times 1.2 = 576 \text{ joules}$$

REMEMBER – Start with the equation and finish with the unit.

Whenever a force moves an object, work is done and energy is transferred from one form into another.

work done = energy transferred

LEARN THIS EQUATION.

Work and energy are both measured in joules (J).

When work is done against frictional forces, most energy is transferred as heat. This is why rubbing your hands together makes them warm.

Elastic potential energy

An elastic object can be stretched, squashed, twisted or bent. But it goes back to the way it was when the force is taken away.

Elastic potential energy is the energy stored in an elastic object when a force does work to change the shape of the object.

CHECK YOUR UNDERSTANDING

1. A motorcyclist is travelling at 18 m/s. Write down three factors that could affect the stopping distance of the motorcyclist.
2. A shopper pushes a loaded trolley 20 m using a constant force of 50 N. How much work does the shopper do?
3. What type of energy does a stretched rubber band store?

The Earth and beyond

The Solar System

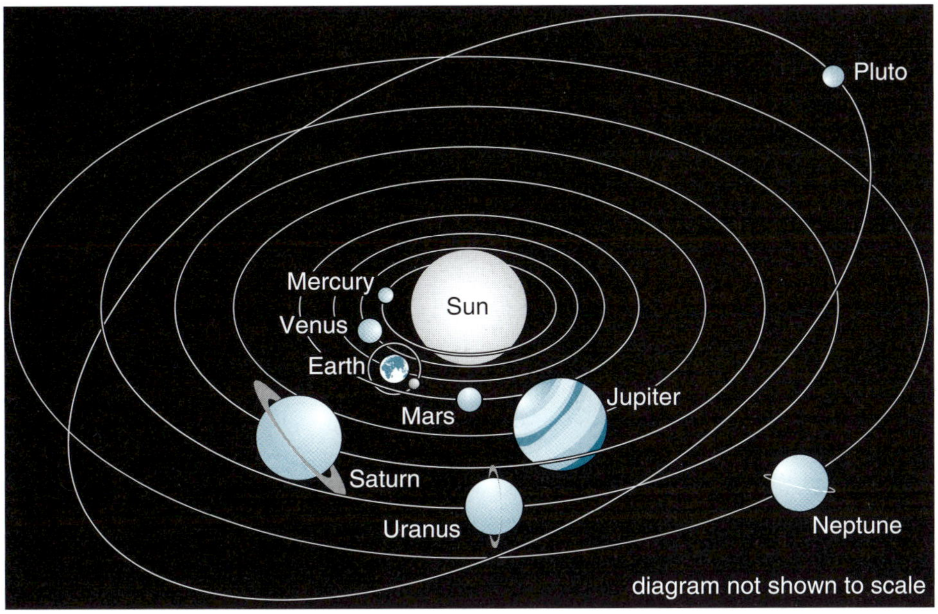

Figure 11.9 The Solar System

The orbits of the planets are elliptical. This means they are like a squashed circle with the Sun close to the centre.

The further a planet is away from the Sun, the longer it takes to complete one orbit.

An **orbit** is the path taken by one object going around another object. For example, the path taken by the Earth around the Sun.

Comets

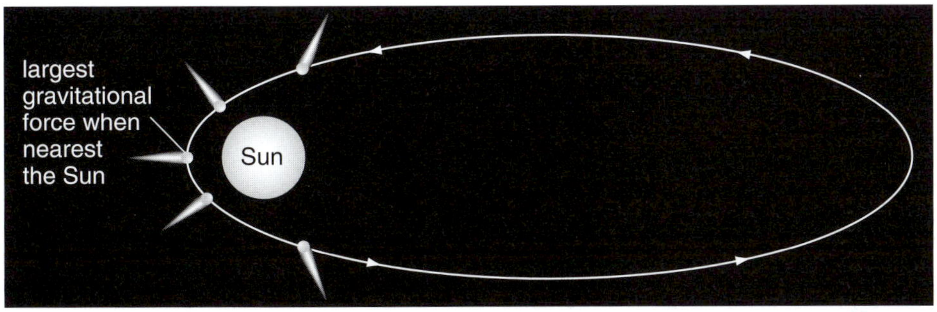

Figure 11.10 The path of a comet

A comet has a very elliptical orbit, so most of the time it will be a huge distance from the Sun. A comet is only seen when its orbit passes close to the Sun.

Gravity in space

The gravitational pull of the Sun affects all objects in the Solar System. The force of gravity between the Sun and a planet together with the high speed of the planet keeps the planet in its orbit.

As the distance between objects gets bigger, the gravitational pull between them gets smaller. When the distance between a rocket and the Earth doubles, the force of gravity goes down by more than half.

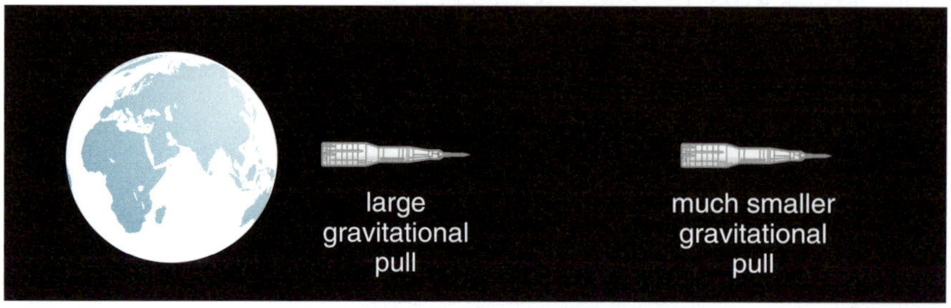

Figure 11.11

Satellites

A **satellite** is an object that orbits a planet. The further a satellite is from the Earth, the longer it takes to complete an orbit. To stay in a particular orbit, a satellite must have the right speed.

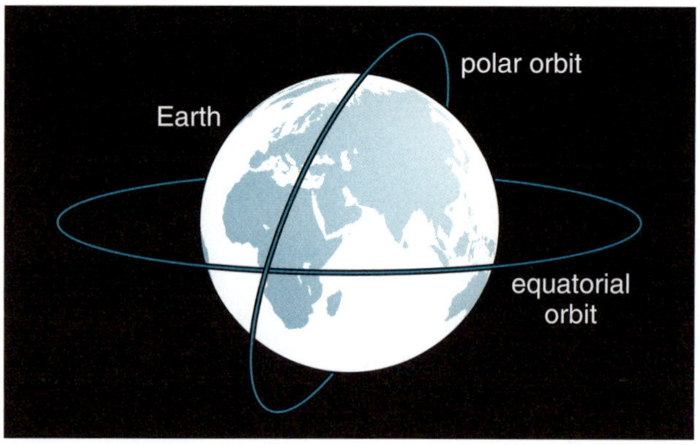

Figure 11.12 Satellite orbits

Satellites in low **polar orbits** are used to monitor the Earth's surface. Such things as forest fires and troop movements can be watched and monitored. Because the Earth spins, the whole of its surface can be scanned each day.

Satellites in high **equatorial orbits** are often used for communications. They are **geostationary orbits**. The satellite takes 24 hours to complete one orbit. This means that the satellite will always be above the same point on the Earth's surface.

CHECK YOUR UNDERSTANDING

1. What force keeps a planet in its orbit around the Sun?
2. Why are communication satellites put into geostationary orbits?
3. Why are spy satellites put into polar orbits?
4. Describe how the gravitational force between the Sun and a comet changes as the comet orbits the Sun.

Outside the Solar System

Figure 11.13

- The Sun is one of the 100 000 million stars in the Milky Way galaxy.
- The distance between stars in a galaxy is often millions of times bigger than the distance between the planets in the Solar System.
- The Universe is made up of at least a billion (a thousand million) galaxies.
- The distance between galaxies is often millions of times bigger than the distance between the stars in a galaxy.

Life on other planets

The search for any form of life in space has not been easy. Unmanned space probes have been sent to Mars and photographs taken of other planets. Experiments have been designed and clues looked for.

The presence of water

Recent photographs of Europa (one of Jupiter's moons) show that its surface could be an icy crust, possibly with water below. There is evidence that there was once water on Mars.

Changes to the atmosphere of a planet

Living things change the atmosphere of a planet. For example, without plants there would be much less oxygen in the Earth's atmosphere.

Fossils in rocks

A meteorite thought to have come from Mars seemed to show the presence of microscopic worm-like structures in the rock.

SETI (Search for Extra-Terrestrial Intelligence)

The search for intelligent radio signals from space has gone on for more than 40 years. So far the project has not found anything.

The life cycle of a star

The birth of a star

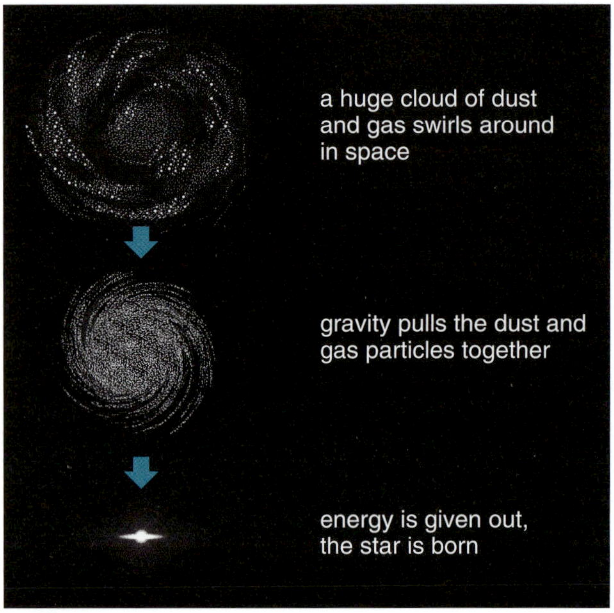

Figure 11.14

Masses too small to become stars join together to form planets and moons.

Stable period of a star

Gravitational forces pulling inwards balance the outward forces caused by the star's high temperature. The Sun is in the stable part of its life cycle.

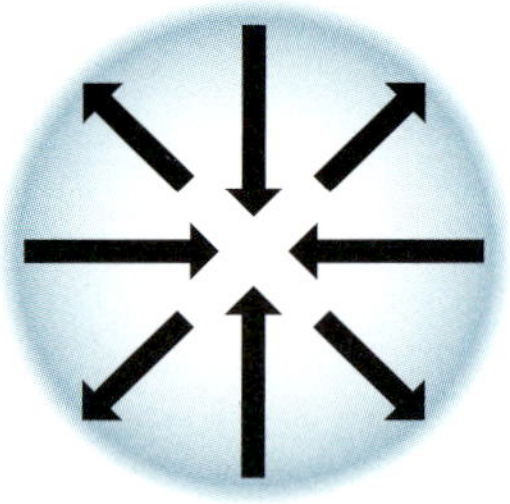

Figure 11.15

The death of a star

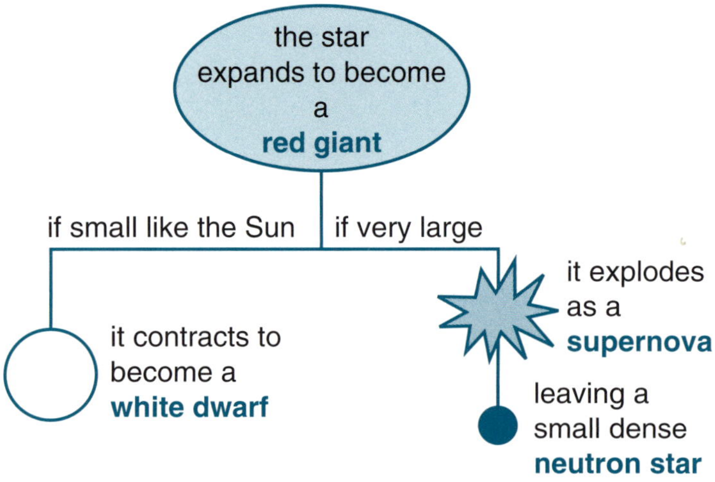

Figure 11.16 The death of a star

CHECK YOUR UNDERSTANDING

❶ What is the Universe?

❷ What will happen to the Sun after it becomes a red giant?

❸ What is left after a supernova explosion?

❹ What clues are there that life may once have existed on Mars?

Topic Test

1 The graph shows how the distance travelled by a person changes with time during a short journey.

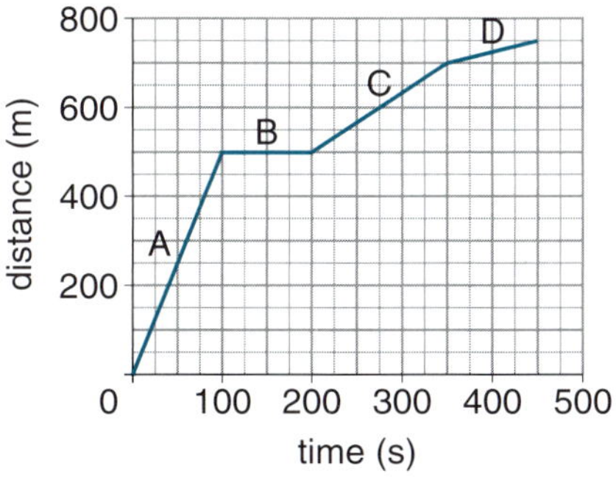

a) What distance was the journey?

b) How long did the journey take?

c) During which part of the journey was the person moving the slowest?

d) How can you tell from the graph that during the journey the person stopped?

(4 marks)

2 How much work is done by a weightlifter lifting a weight of 2500 newtons to a height of 2.1 metres above the ground? (3 marks)

3 Give three different uses for a satellite. (3 marks)

4 Copy and complete each sentence by choosing the correct word from the box. Each word may be used once or not at all.

friction	galaxy	gas	gravity	liquid	planets	stars	Universe

The Sun is in the _____________ called the Milky Way. The _____________ is made up of at least a billion galaxies. _____________ like the Sun were formed from clouds of dust and _____________ pulled together by _____________. (5 marks)

5 How would an analysis of the atmosphere of a planet give evidence that life may exist or have existed at some time in the past on the planet? (2 marks)

6 Describe how the SETI project is trying to gain evidence for the existence of intelligent life on other planets. (2 marks)

7 Describe the motion of the runner whose velocity–time graph is shown here.

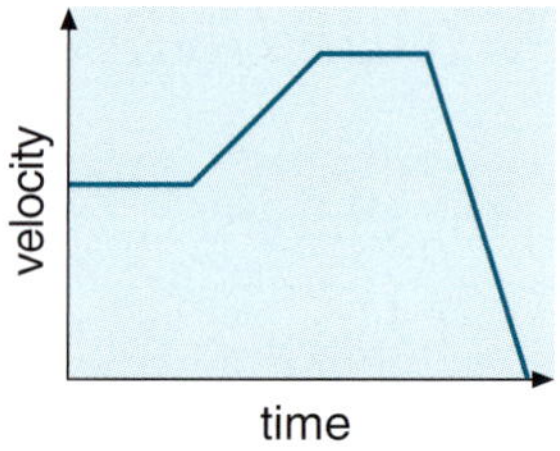

(4 marks)

8 Why do cyclists on a level road slow down when they stop pedalling? (2 marks)

9 State and explain the effect on the stopping distance of a car when the road is wet. (2 marks)

10 How much would a person of mass 70 kg weigh on the Moon? (Gravitational field strength on the Moon = 1.6 N/kg.) (3 marks)

Total: 30 marks

WAVES AND RADIATION

MODULE 12

Waves and waveforms

Energy can be transferred by **waves**. A stone falling on water causes ripples to move across the water surface. The water particles only move a small distance around their starting position, but the energy travels outwards to another place.

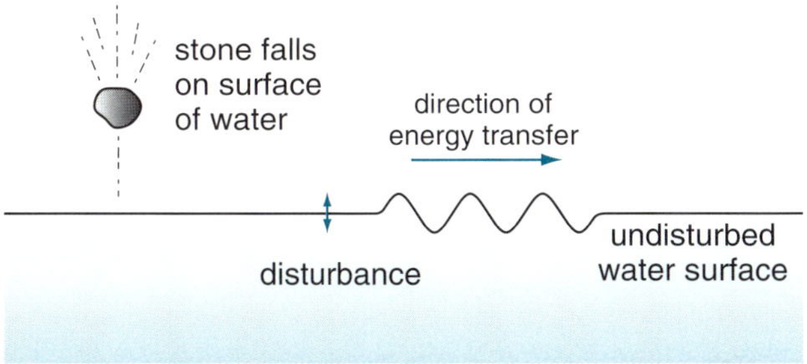

Figure 12.1 Ripples on a water surface

Waves can also be seen on a rope

Figure 12.2 Waves on a rope

or a spring.

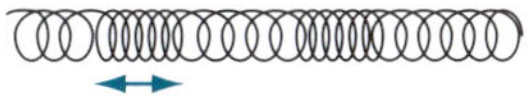

Figure 12.3 Waves in a 'slinky' spring

Sound waves can only travel through a solid, liquid or gas.

Light and radio waves can travel through a vacuum.

Types of wave

Transverse wave – the disturbance is at 90° to the direction the wave travels (direction of energy transfer).

Examples: Light waves and water waves.

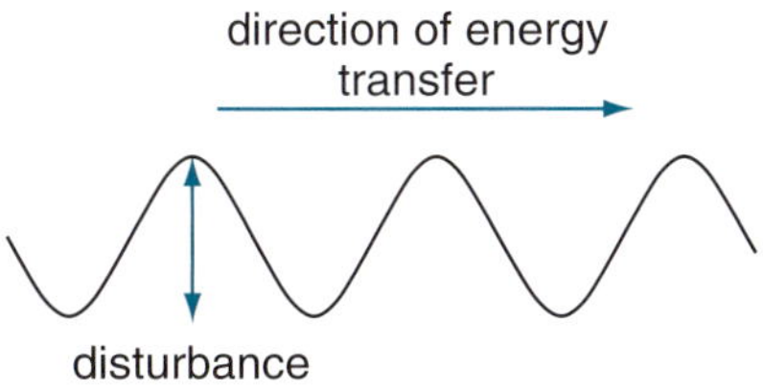

Figure 12.4 A transverse wave

Longitudinal wave – the disturbance is along (parallel to) the direction the wave travels (direction of energy transfer).

Examples: Waves along a slinky spring and sound waves in the air.

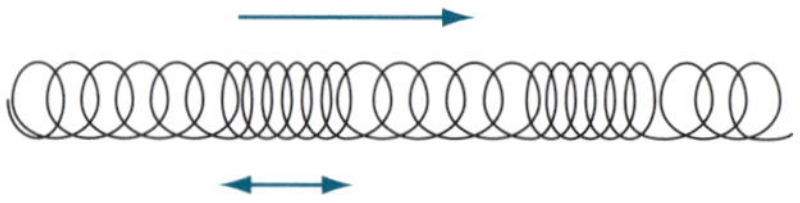

Figure 12.5 A longitudinal wave

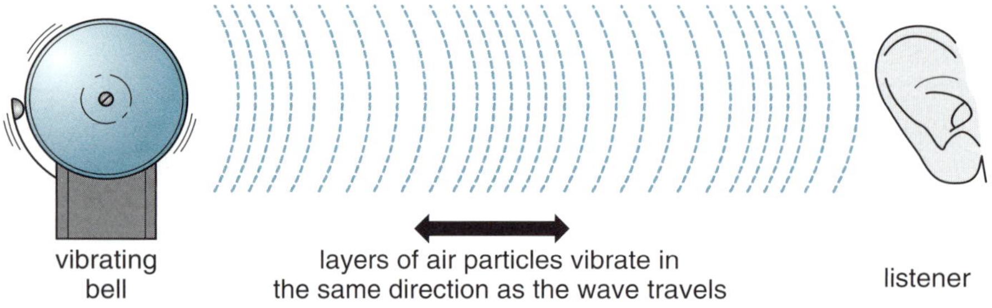

Figure 12.6 A longitudinal sound wave

Waveform features

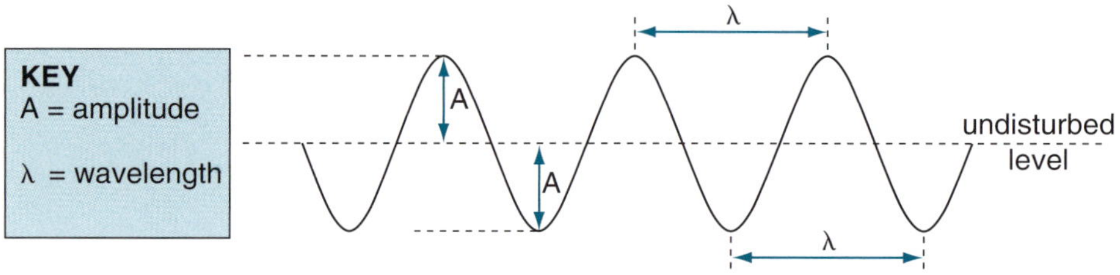

Figure 12.7 The features of a wave

Some wave definitions:

Frequency = the number of waves passing a particular point in 1 second, or the number of waves produced by a source in 1 second.

Wavelength = the distance from one point on a wave to the same point on the next wave.

Amplitude = the maximum disturbance of the wave.

Frequency is measured in **hertz** (Hz).

The wave equation relates the wave speed to the frequency and wavelength.

wave speed = frequency × wavelength
wave speed in metres/second (m/s)
frequency in hertz (Hz)
wavelength in metres (m)

LEARN THIS EQUATION and the UNITS.

Example: Calculate the speed of a water wave that has a wavelength of 3 m and a frequency of 20 Hz.

wave speed = frequency × wavelength
= 20 × 3 = 60 m/s

CHECK YOUR UNDERSTANDING

❶ What is the difference between a transverse and a longitudinal wave? Give one example for each type of wave.

❷ Which one of the following waves cannot travel through a vacuum?

light radio sound

❸ A set of 30 regular water waves occupies a distance of 360 cm. The 30 waves take 5 s to pass a float.

a) What is the wavelength of the waves?

b) What is the frequency of the waves?

c) Use the wave equation to calculate the speed of the waves.

Behaviour of waves

Reflection

Water waves, like light waves, can be **reflected** by a surface.

Waves on a rope or spring can also be reflected.

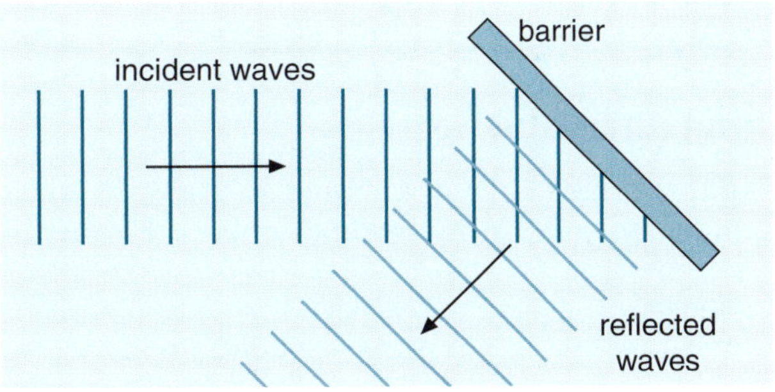

Figure 12.8 Reflection of water waves

Refraction

Water waves move faster across deep water than shallow water. So water waves moving across water of different depths change speed. This will usually make the waves change direction. When the waves change direction, they have been **refracted**.

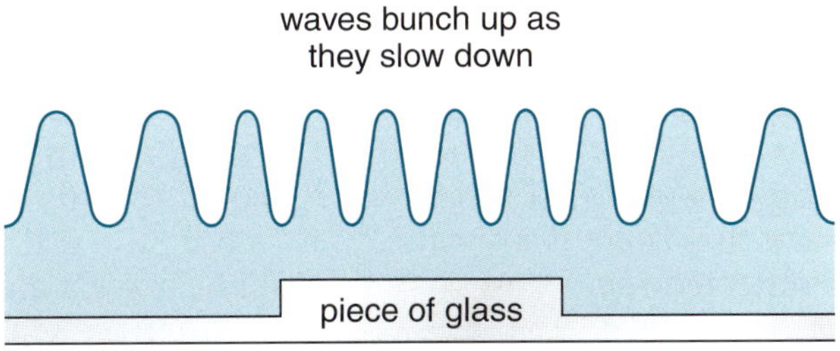

Figure 12.9 Change in speed of water waves

Waves are not refracted when the direction of travel of the waves is along a normal.

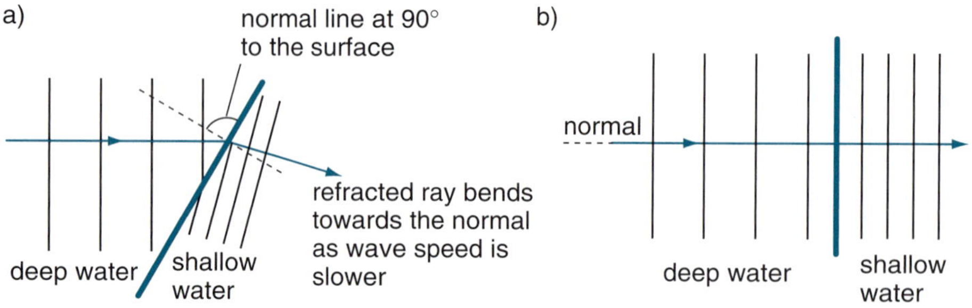

Figure 12.10

Sound and light waves change speed when they travel into a different medium. The change in the speed causes a change in direction (refraction) provided the waves are not travelling along a normal.

CHECK YOUR UNDERSTANDING

1. Do water waves travel faster or slower when they reach deeper water?
2. What happens to the wavelength of water waves as they travel from shallow to deeper water?
3. A light wave travelling from air into glass does not refract. Why?

Total internal reflection

Light travels faster in air than in glass, Perspex or water. The light ray bends away from the normal when it enters the air.

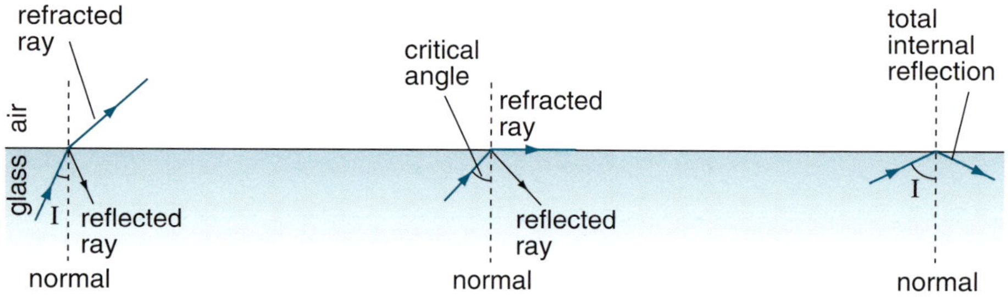

The critical angle for glass = 42°.

At small angles (I) most light is refracted into the air. There is a weak reflected ray.

When the angle I is equal to the **critical angle**, the refracted light travels along the edge of the glass.

When the angle I is bigger than the critical angle, all the light is reflected back into the glass. This is called **total internal reflection**.

Figure 12.11 Total internal reflection

Optical fibres pass light into places that are difficult to reach. In an optical fibre, light can travel to, then back from, the object to be seen by the observer. This idea is used in an **endoscope**, an instrument used by doctors to see inside a patient's body.

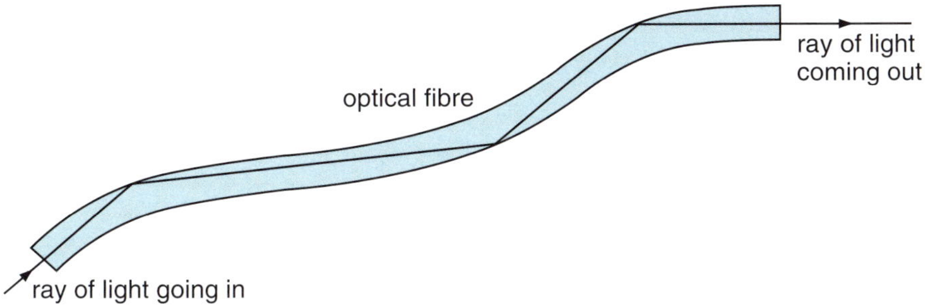

Figure 12.12 Passage of light through an optical fibre

Diffraction

All waves can **diffract** from the edge of an obstacle or through a gap. This spreading out of waves is easily seen in water waves, because the wavelength of the waves often matches the size of the gap.

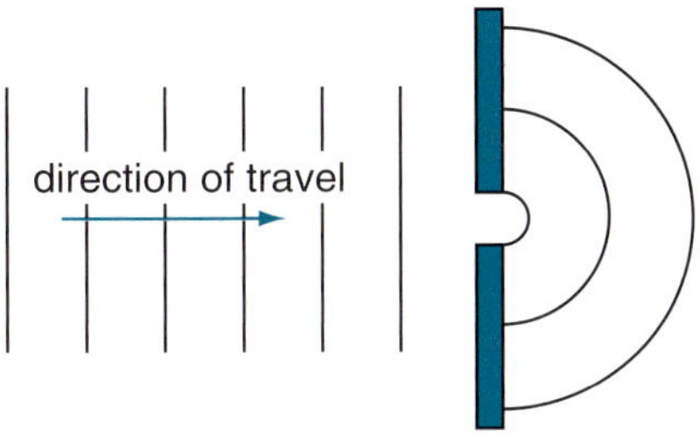

Figure 12.13 Diffraction of waves at a gap

Diffraction allows sound waves to spread out from doorways. Light waves are diffracted but not easily because their wavelengths are very small.

Radio waves can also be diffracted.

The hill diffracts the waves to the house aerial.

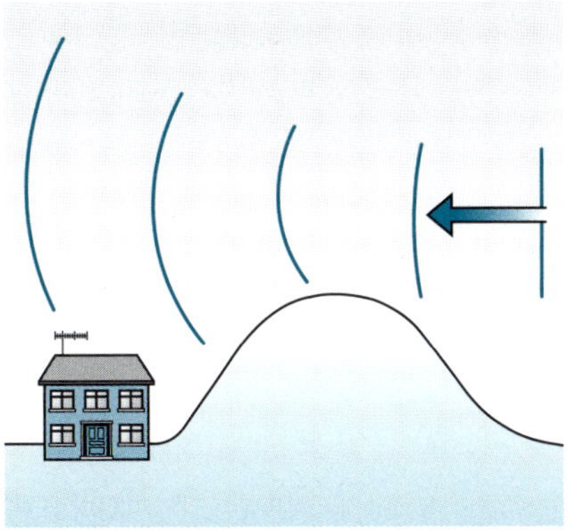

Figure 12.14 Diffraction of radio waves

CHECK YOUR UNDERSTANDING

1. The aerial on a house in a valley can pick up long wavelength radio waves but not short wavelength radio waves. Explain why.
2. How does the intensity of the reflected light in a glass block change as the angle between the incident ray and normal increases?

Electromagnetic waves

- Electromagnetic waves are transverse waves with a range of wavelengths and frequency.
- They all travel through space (a vacuum) at the same speed, 300 000 000 m/s.
- Different wavelengths are transmitted, reflected or absorbed differently by different materials and surfaces.
- When absorbed, electromagnetic waves make the absorbing material hotter.
- When absorbed by a conductor they may produce an alternating current (a.c.) with the same frequency as the waves.

highest frequency
shortest wavelength

Type of wave	Uses	Effect on living cells
Gamma rays	Kill harmful bacteria in food Sterilise medical instruments Kill cancer cells	High doses kill normal cells. Low doses easily cause cell change leading to cancer.
X-rays	Make shadow pictures of bones (absorbed by bone and metals)	Pass through soft tissue but some will be absorbed. High doses will kill cells but low doses may cause cancer.
Ultraviolet rays	Sunbeds Labels for security coding Fluorescent lamps – coating absorbs UV and emits energy as light	Can pass through skin to reach deeper tissue; this may lead to skin cancer. (Pale skin absorbs less UV than darker skin so more UV reaches the deeper tissue.)
Visible light	Seeing Optical fibre instruments and communications	
Infrared rays	Cooking – grills and toasters Radiant heaters Optical fibre communications Remote controls for VCR and TV sets	Absorbed by the skin and felt as heat; may cause pale skin to go red.
Microwaves	Satellite and mobile phone communications Cooking – strongly absorbed by water molecules causing a heating effect	All living cells contain water, so are killed or damaged if exposed to microwaves.
Radio waves	Transmit TV and radio programmes	

lowest frequency
longest wavelength

Figure 12.15 The electromagnetic spectrum

Communication and electromagnetic waves

Radio waves send signals to different places on the Earth.

- Long wavelength radio waves reflect from the ionised layers in the upper atmosphere.
- Short wavelength radio waves carry TV signals across small distances on the Earth.
- Microwaves are very short wavelength radio waves which penetrate the atmosphere to satellites.

  These waves are then sent back to Earth.

  This system is used by mobile phone networks and satellite TV stations.

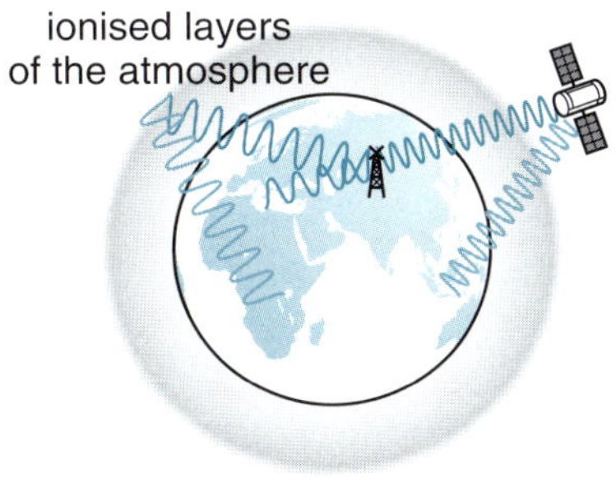

Figure 12.16 Using radio waves for communication

Signals

Sound waves can be changed to electrical signals.

These signals can follow exactly the pattern of the sound wave, given by music or speech. This is an **analogue signal**.

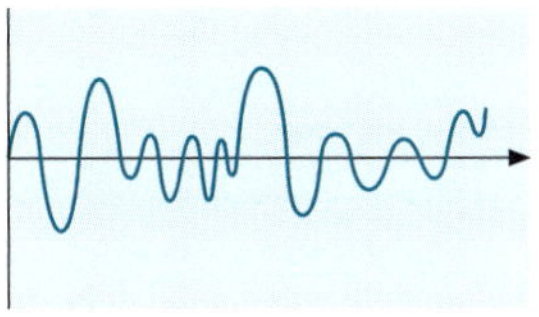

Figure 12.17 An analogue signal

Signals coded as a series of 'on or off' pulses are **digital signals**.

Digital signals are less likely to change the information they carry and they transfer more information in a given time than analogue signals.

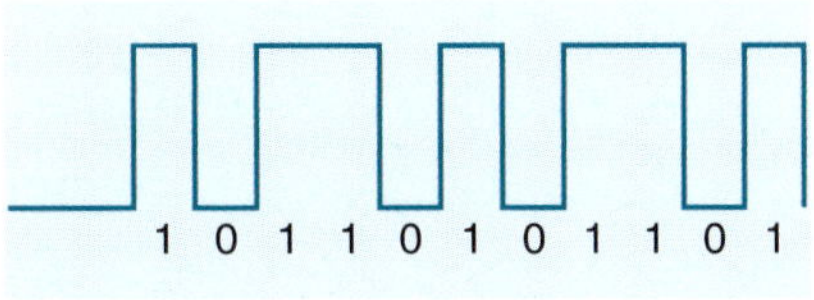

Figure 12.18 A digital signal

- Signals can be carried directly as a varying electric current in copper cables. Or the electrical signal can be transferred to a radio wave and carried in a radio transmission.
- A digital signal can be changed to light or infrared pulses and carried in an **optical fibre**.
- An optical fibre can carry more telephone messages at the same time than a copper cable of the same diameter.
- The strength of light signals does not weaken as quickly as that of electrical signals.

CHECK YOUR UNDERSTANDING

❶ List these electromagnetic waves in order of decreasing wavelength:
gamma rays, infrared, light, microwaves, radio waves, ultraviolet, X-rays

❷ a) When do some substances fluoresce?
b) Where is fluorescence used?

❸ Match the correct labels to each stage in this telephone system. Use each label once.
digital light signal, satellite, electrical signal,
microwaves, optical fibre cable, copper cable

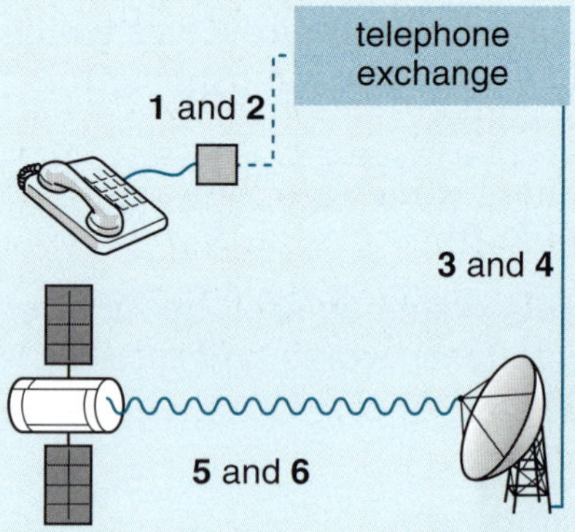

Waves travelling through matter

Sound and ultrasound

Sound waves, produced when objects vibrate, can be shown as an oscilloscope trace.

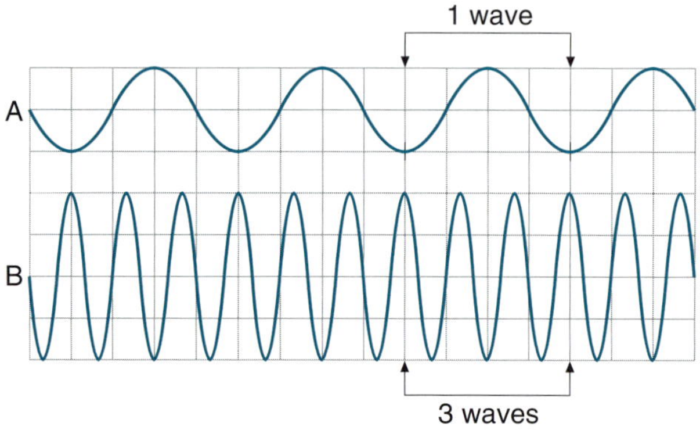

The amplitude of wave B is twice the amplitude of wave A.

Wave B has three complete waves in the time it takes wave A to complete one wave. Wave B has three times the frequency of wave A.

Figure 12.19 The trace on an oscilloscope screen

The human ear can detect sound waves with frequencies from 20 Hz–20 000 Hz (20 kHz).

Sound waves with frequencies above the upper limit of human hearing are called **ultrasound** or **ultrasonic waves**.

Ultrasonic waves reflect and refract at different boundaries in materials.

At low intensities, ultrasound waves do not harm living cells. In medicine, reflected ultrasonic waves are used to produce an image of an unborn baby (pre-natal scanning).

In industry, ultrasonic waves are used for quality control (e.g. finding cracks in metal structures) and cleaning (e.g. a watch mechanism).

Seismic waves

Earthquakes cause shock waves in the Earth causing the Earth's crust to move. The shock waves can travel through the interior of the Earth. These are seismic waves and they are recorded on seismographs. The arrival of the reflected and refracted waves after the initial disturbance gives the position of the earthquake.

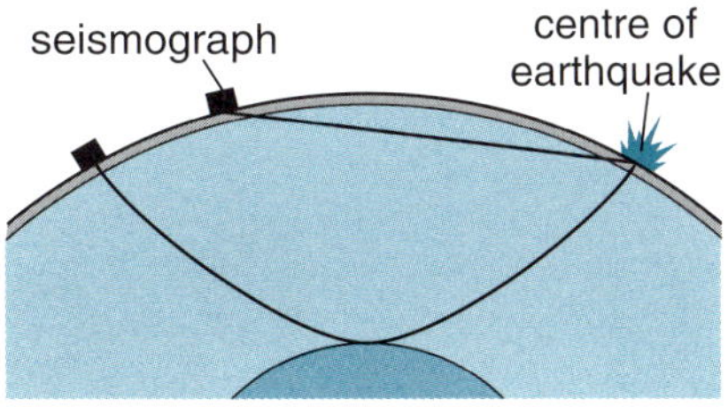

Figure 12.20

These signals can also give information about the Earth's layered structure.

CHECK YOUR UNDERSTANDING

1. a) Wave Y has an amplitude of 3 cm. What is the amplitude of wave X?
   b) Wave X has a frequency of 40 Hz. What is the frequency of wave Y?

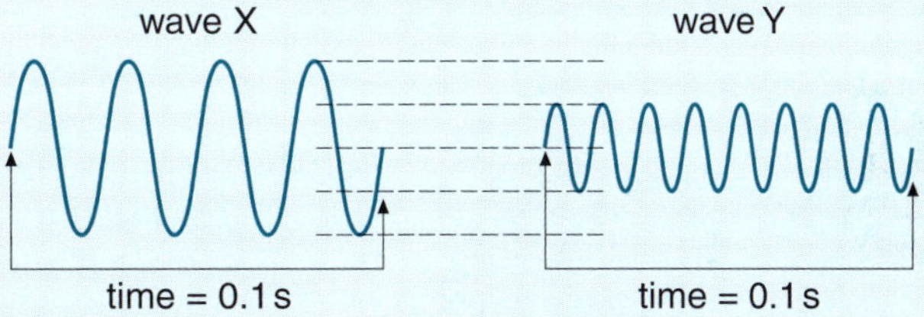

2. Why is it better to use ultrasonic waves than X-rays to obtain an image of an unborn baby?

Structure of the atom

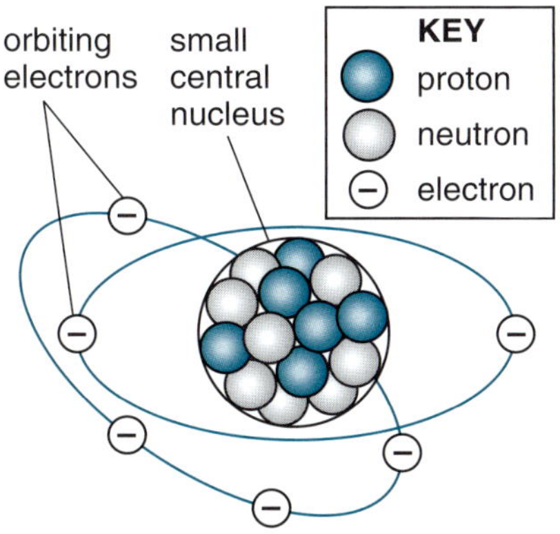

Figure 12.21 The basic structure of an atom

In an electrically neutral atom, the number of electrons = the number of protons.

The relative masses and charges of protons, neutrons and electrons are shown in the table.

Particle	Mass	Charge
Proton	1	+1
Neutron	1	0
Electron	0	−1

LEARN this table.

All atoms of the same element have the same number of protons.

Protons and neutrons are both called **nucleons**.

The total number of protons and neutrons in the nucleus is the **nucleon number** or **mass number**.

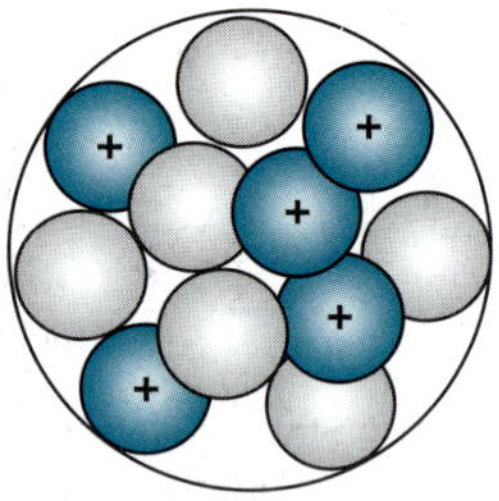

Figure 12.22 The nucleus has five protons and six neutrons.
Mass number = 11

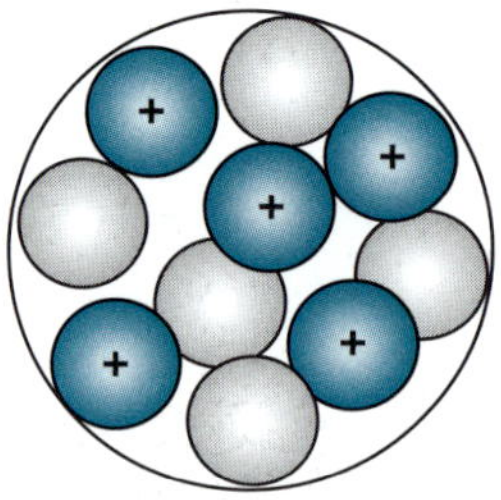

Figure 12.23 This nucleus has five protons and five neutrons.
Mass number = 10

An **isotope** of the same element has the same number of protons but a different number of neutrons. Figures 12.22 and 12.23 are isotopes of the same element.

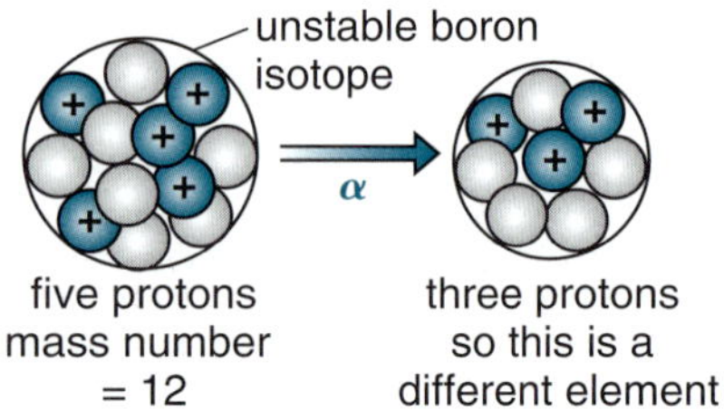

Figure 12.24 The radioactive decay of a boron nucleus

A **radioisotope** is another form of the same element where the nucleus is **unstable**. When such a nucleus decays by emitting an α or β particle, then the structure of the nucleus changes. The new atomic nucleus has a different number of protons.

CHECK YOUR UNDERSTANDING

1 Which of the nuclei below:

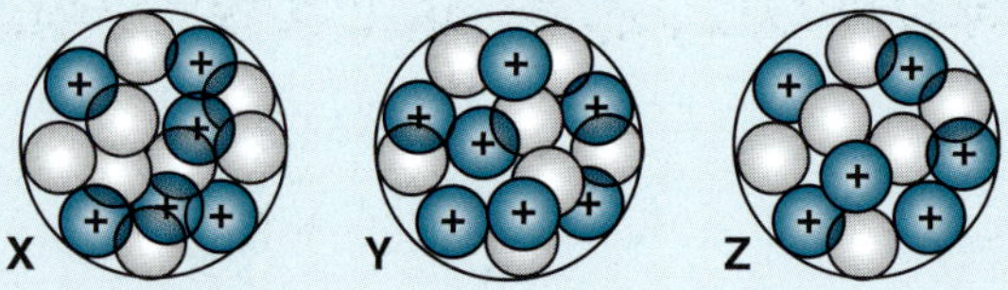

a) have the same mass number?

b) are isotopes of the same element?

c) has the highest number of protons?

Radioactivity

Radioactive substances give out radiation in the form of particles and gamma rays. They do this at random, regardless of what is done to them. Atomic nuclei of the radioactive substances are unstable. When they give out radioactivity, they change their structure. These substances are called radioisotopes or radionuclides, and the process is called radioactive decay.

The types of radiation are summarised in this table.

Type of radiation	Penetrating ability	Ionising ability
alpha α	Low. Stopped by paper or 5 cm of air	High
beta β	Fair. Stopped by a few mm of metal	Fair
gamma γ	High. Reduced by 10 cm of lead or several metres of concrete	Low

Background radiation is the low level, continual radiation received mostly from natural sources e.g. rocks, cosmic rays from space, building materials and food. A very small part of the background radiation is due to man-made sources e.g. nuclear reactors, medical X-rays.

The dangers of radiation are reduced by thick shielding. The more dense and wider the absorbing material, the less radiation is transferred. Radioactivity can kill living things. It ionises the electrically neutral atoms or molecules in living cells. The chemical structures and functions of the cells change. The cells die or they stop working normally, causing cancers.

- Outside the body – alpha radiation is the least dangerous (unlikely to reach living cells).
  – beta and gamma radiation are the most dangerous (can reach cells of organs and be absorbed).
- Inside the body – alpha radiation is the most dangerous (strongly absorbed by cells).
  – beta and gamma radiation are less dangerous (may pass through cells without being absorbed).

A badge detecting radiation is often worn by people working with radioactive materials. The badge has a photographic film. This records the type and amount of radiation received. The darker the film after developing, the more radiation the worker has been exposed to.

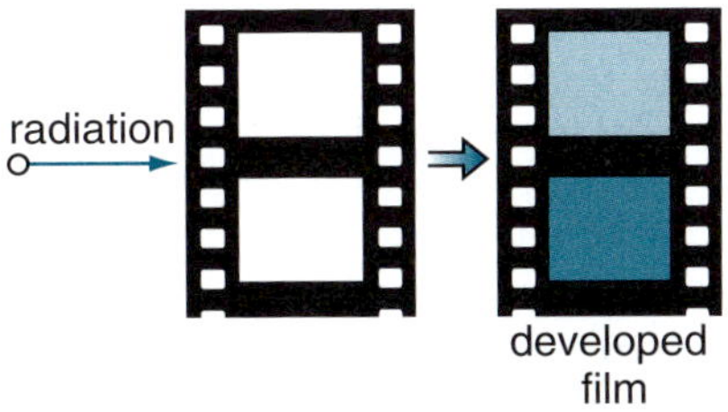

Figure 12.25 The badge film detects radiation

Radioactive **half-life** is the time taken for the count rate to fall to half the starting value, or for half the number of unstable nuclei from the parent atoms to decay. Very active samples have short half-lives as they lose their radioactivity at a fast rate.

For Figure 12.26, the time taken for the count rate to go from 1200 to 600 or from 600 to 300 is 10 hours.

The half-life for this sample is 10 hours.

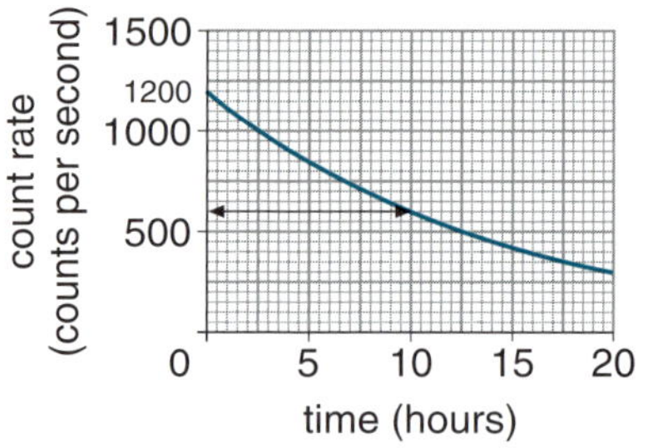

Figure 12.26 A graph showing radioactive decay

You need to be able to find the half-life from a graph like this.

Using radiation

1 The ability of radiation to penetrate or be absorbed by different materials is used to control the thickness of materials. When the material passes too much radiation then it is too thin, and the rollers are adjusted.

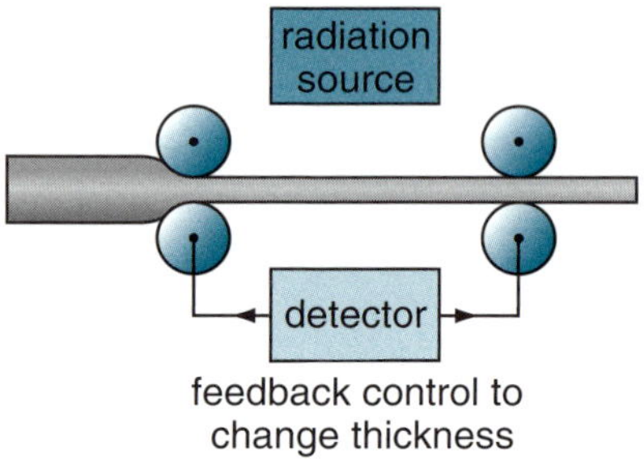

Figure 12.27 Controlling the thickness of materials

2 The destruction of living cells by radiation is used in:

- Sterilising medical instruments, as the radiation kills the harmful organisms.
- Prolonging the shelf life of food, by reducing the number of food-spoiling bacteria.
- In radiotherapy, for killing cancer cells.

3 Radiation can be used to date ancient materials, including rocks. The older the materials, the less radiation they emit.

CHECK YOUR UNDERSTANDING

❶ Name a natural source of background radiation.

❷ Which type of radiation is most dangerous

a) when the source is some distance outside the body?

b) when the source is inside the body?

❸ The table shows the count rate from a radioactive sample decreasing with time. What is the half-life of the sample?

Counts/minute	960	760	600	480	380	300	240	190
Time in days	0	1	2	3	4	5	6	7

Topic Test

1 Water waves of wavelength 2.5 cm pass a fixed marker at the rate of 3 waves per second.
  a) What is the frequency of the waves in hertz?
  b) Calculate the speed of these waves. Write down the equation you will use and show your working out. (5 marks)

2 Write these waves in the correct columns of a table like the one below. Some waves are in more than one column.

waves on a stretched spring, light waves, sound waves, radio waves, surface water waves

Longitudinal waves	Transverse waves	Waves crossing a vacuum

(3 marks)

3 Water waves cross at an angle to the boundary from shallow to deep water.
Describe what changes, if any, occur to the following:
a) direction, b) speed, c) wavelength.

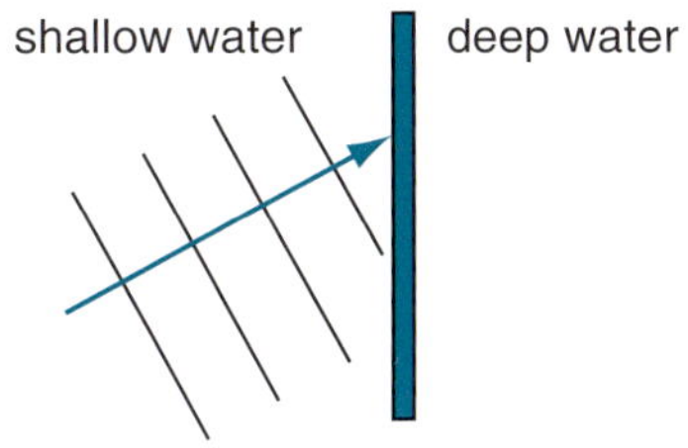

(3 marks)

4 Which waves from the electromagnetic spectrum can be used for the following?
  a) cooking
  b) providing shadow pictures of bones
  c) TV remote control units
  d) night-time photography
  e) obtaining a suntan (5 marks)

5 a) Define ultrasound.
  b) Give two uses of ultrasonic waves. (3 marks)

6 a) What is meant by the term background radiation?

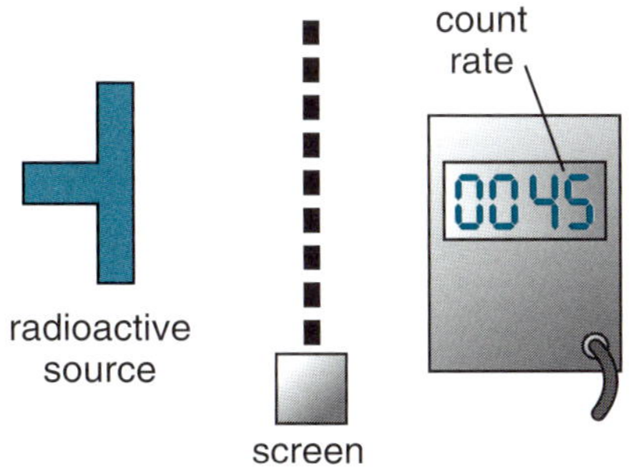

Screen	Count rate in counts/minute
No absorber	542
Thin paper	268
5 mm metal	271
10 mm lead	45

A radioactive source is placed behind different screens. The radiation detected at the other side of each screen is shown in the table. The background radiation during this investigation is 12 counts/minute.
  b) Why is the background radiation measured?
  c) What type(s) of radiation is the source emitting? Explain the reason for your choice. (5 marks)

Total: 24 marks

Physics examination questions and model answers

1 a) Different energy sources can be used to generate electricity.

Wind is an energy source	Coal, a fossil fuel, is an energy source
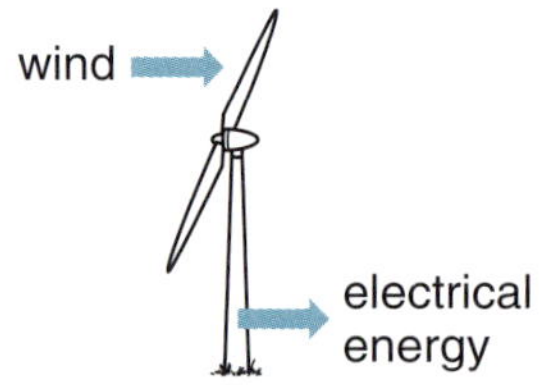	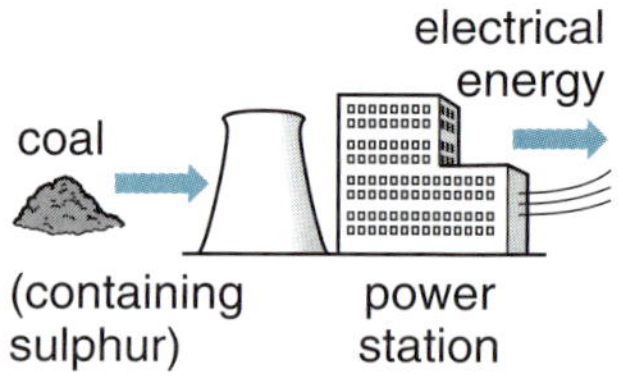
This wind turbine generates 1 MW (1 MW = 1000 kW)	This coal-fired power station generates 1000 MW
Electricity demand in the UK can be 48 000 MW	

Give one advantage and one disadvantage (other than cost) of using each energy source to generate electricity in the UK. *(4 marks)*

Advantage	Disadvantage
Using wind *It uses a renewable energy resource.*	Using wind *The amount of electricity generated depends on the wind strength. So it is not reliable.*
Using coal *Only 48 power stations are needed to generate the electricity compared to 48 000 wind turbines.*	Using coal *Burning coal releases sulphur dioxide which causes acid rain, OR releases carbon dioxide which causes increase in greenhouse effect.*

Write one answer in each space. Ensure the answers do not refer to cost. Make sure you make use of the information given in the question.

b) List A gives the names of three electrical devices.
List B gives the type of useful energy transferred.

Draw a straight line from each electrical device in List A to the useful energy it transfers in List B. *(2 marks)*

List A (Electrical device) — List B (Useful energy transferred)

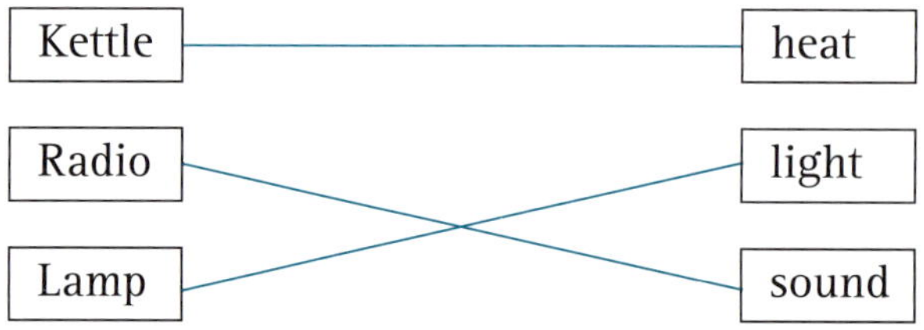

Draw only three lines. You should never draw more than one line to or from a box.

2 a) The table gives information about some planets.

Planet	Gravitational field strength in N/kg	Diameter of planet in thousands of km	Time for one orbit around the Sun in years
Mercury	4	4.9	0.2
Venus	9	12.0	0.6
Earth	10	12.8	1.0
Jupiter	26	143.0	12.0

(i) Write down the equation that links gravitational field strength, mass and weight. *(1 mark)*

weight = mass × gravitational field strength

Always write the equation using words not symbols. Correct symbols are acceptable, but it is easy to make a mistake.

(ii) An astronaut has a mass of 75 kg. Calculate the weight of the astronaut on Venus. Show clearly how you worked out your answer. *(2 marks)*

weight = 75 × 9 = 675

weight of astronaut = 675 newtons

Always show the steps in your working out. If you make a mistake with the arithmetic you can still get a mark, or marks, for a correct method.

(iii) Why would the astronaut weigh more on the Earth than on Venus? *(1 mark)*

The gravitational field strength is higher on Earth than it is on Venus.

Use the information given in the table. The question asks for a comparison to be made. The answer must mention both Earth and Venus.

(iv) Jupiter is further from the Sun than Mercury, Venus and Earth. What evidence in the table shows this? *(1 mark)*

The time taken by Jupiter to orbit once around the Sun is longer than the time taken by the other planets.

The correct information, in this case orbit time, must be used. A comparison of diameter or gravitational field strength would not gain the mark.

b) Explain briefly how stars like the Sun are thought to have been formed. *(2 marks)*

The Sun was formed from a gigantic cloud of dust and gas (mainly hydrogen). This cloud was pulled inwards and squashed into a smaller volume by the force of gravity.

The question is worth two marks, so make sure you make two clear statements. The two statements should be in the correct order.

Answers

Module 1 – Humans as organisms

Check your understanding

Page 4

1 We are always breathing more oxygen into the lungs and the blood is constantly carrying it away from the alveolus surface.
2 a) Glucose
b) Amylase
3 a) Fatty acids and glycerol
b) Lipase
4 Hydrochloric acids and protease enzymes.
5 In the villi in the small intestine.

Page 6

1 glucose + oxygen ⟶ carbon dioxide + water + energy
2 Without oxygen.
3 When the body still needs oxygen after exercise to respire away lactic acid.
4 Movement, keeping warm, making large molecules from small ones.

Page 10

1 Platelets
2 Any three from:
- Wash regularly.
- Cook food well.
- Cover food up.
- Store food in a cool place.
- Keep away from people with infectious diseases.

Topic test

1 a) tail; nucleus
b) haemoglobin; oxygen
c) shorten/contract (1 mark each)
2 a) Controls cell activity. (1 mark)
b) Where chemical reactions take place. (1 mark)
c) Determines what leaves and enters the cell. (1 mark)
3 Movement of molecules of gas or substance in solution (1 mark)
from region of higher concentration (1 mark)
to a region of lower concentration (1 mark)
4 Large differences in concentration (1 mark)
Large surface area of membrane (1 mark)
5 a) villi (1 mark)
b) alveoli (1 mark)
6

Insoluble food	Enzyme	Soluble food
starch	amylase	sugars
proteins	protease	amino acids
fats	lipase	fatty acids and glycerol

(1 mark each)

7 neutralises; stomach; enzymes; fats; surface area (1 mark each)
8 Ribs move out. (1 mark)
Diaphragm flattens. (1 mark)
9 Both sides contract at once. (1 mark)
Atria contract and push blood into ventricles. (1 mark)
Valves close. (1 mark)
Ventricles contract and push blood in two directions. (1 mark)
Ventricles relax and artery valves close. (1 mark)
10 carbon dioxide; soluble food; urea (1 mark each)
11 Weakened micro-organism injected into blood. (1 mark)
White blood cells make antibodies. (1 mark)
Body can make these again quickly (1 mark)
if real micro-organism gets in. (1 mark)
Antibodies ready to destroy them. (1 mark)

Answers

Module 2 – Maintenance of life

Check your understanding

Page 15

1 Hot, dry and windy
2 Because desert plants live in extremely dry conditions and so need to minimise the amount of transpiration and conserve the water in the plant.

Page 17

1 Light = eye, sound = ear, taste = tongue
2 Goalkeeper has to think about it and decide what to do.
3 Any two from:
blinking, swallowing, knee jerk reflex, iris changing size.

Page 19

1 Carbon dioxide, urea
2 Insulin

Page 20

1 A chemical that affects the way the body chemistry works or the way we feel.
2 Any two from:
alcohol, solvents, tobacco.
3 Damage to brain and liver, coma, lack of self control.

Page 21

1 Some cells make mucus, others are lined with cilia.
2 Emphysema, bronchitis, lung cancer

Topic test

1 a) Controls what enters and leaves the cell.
b) Makes the cell rigid.
c) Absorbs light to make food.
(1 mark each)
2 It gives it a large surface area to absorb more water. (1 mark)
3 chlorophyll; starch; leaves; stomata
(1 mark each)
4 carbon dioxide + water ⟶ glucose + oxygen (1 mark)
5 Temperature, carbon dioxide, light
(1 mark each)
6 xylem; sugars (1 mark each)
7 Any two from:
as weedkiller; to root cuttings;
affect ripening of fruit. (1 mark each)
8

Receptors	What it detects
skin	touch and pressure changes
tongue	chemicals so that we can taste and smell

(2 marks)
9 a) Fine focuses light on retina. (1 mark)
b) Changes size of pupil/controls amount of light into the eye. (1 mark)
10 Temperature, ion content, water content.
(1 mark each)
11 Lowers blood sugar level. (1 mark)
12 Feeling ill when drug is not taken. (1 mark)

Module 3 – Environment

Check your understanding

Page 24

1 Thick fur, a thick layer of fat and although they are large, they have a compact body shape with small ears.
2 Food, nesting/sleeping sites, water and a mate.
3 The trees keep out the light (and also take up much of the water).
4 a) Allowed the plants to grow tall.
b) The foxes (predators) were deprived of an important food source so fox numbers reduced and they competed for alternative food sources.

Answers

5 The tree would not be able to photosynthesise so there would be less sugar transported to the fruit so fewer apples produced.

Page 25

1 Some of the light is reflected from the leaf surfaces.
2 a) temperature
   b) (muscle activity for) movement
3 There is less energy lost in respiration and maintaining the biomass.
4 Most biomass is in the producers – the plant level.

Page 27

1 If plants are harvested, the mineral ions are removed from the food chain and the ecosystem.
2 The ideal conditions for composting are aerobic (lots of oxygen) and moist – it will be warm from the respiration of the micro-organisms.
3 Photosynthesis.

Page 29

1
- Sulphur dioxide and nitrogen oxides in exhaust fumes.
- Gases dissolve in water droplets in rain clouds.
- Acid rain falls (low pH).
- Acid rain lowers pH of lake water and harms plants, insects and fish or falls on trees and damages leaves
  OR lower pH water taken up by the tree roots, carried up to the leaves and damages them resulting in less photosynthesis.

2 a) global warming, b) sulphur dioxide,
   c) acid rain
3 Burning or composting.
4 Fertilisers, pesticides, sewage or toxic chemicals.
5 There will be less carbon dioxide, sulphur dioxide and nitrogen oxide, therefore global warming and acid rain problems would be reduced.
6 The greenhouse effect causing global warming.

Topic test

1 b), d), e), f), h) (1 mark each)
2 a)–(v), b)–(i), c)–(vi), d)–(iii), e)–(iv), f)–(ii) (6 marks)
3 a) There is competition for food (1 mark) and so there is not enough food to raise more than eight chicks in the wood. (1 mark)
   b) Increased predation; (1 mark) population will fall as robins are eaten by the sparrowhawk. (1 mark)
   c)

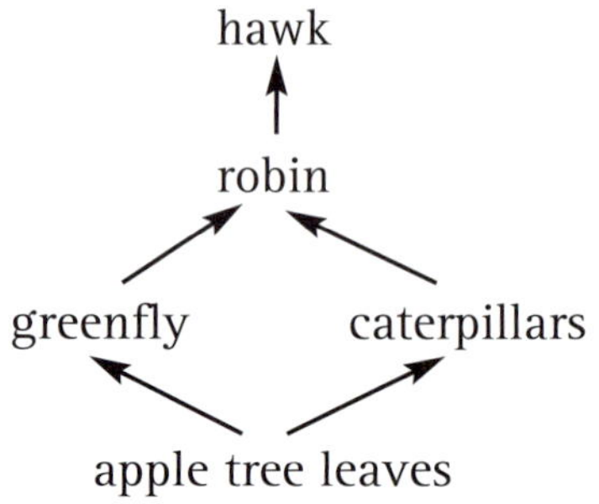

(1 mark for all links correct. 1 mark for the arrows in the correct direction.)

4 a) He turns the heap to aerate it (1 mark) as the decomposing bacteria are aerobic/need oxygen for respiration. (1 mark)
   b) It is kept moist because the bacteria need moisture to survive/for their enzymes to work. (1 mark)
   c) It is covered to keep it warm because the bacteria (1 mark) reproduce faster in warm conditions (1 mark)
   OR the bacterial enzymes (1 mark) function better in warm conditions. (1 mark)
5 a)

SUN —1%→ PLANT —10%→ HERBIVORE —20%→ CARNIVORE

(3 marks)

   b) Any two from: respiration; muscle use in movement; keeping warm. (2 marks)
   c) Restrict movement/keep them in a small pen; warm hut; kill as soon as they have reached adult size. (3 marks)

Answers

Module 4 – Inheritance and selection

Check your understanding

Page 32

1 a) (i) sperm, (ii) eggs/ova
  b) fertilisation
  c) genes
  d) alleles
  e) dominant
2 Different genes; different environmental conditions.
3 Environmental factors/lifestyle differences – they may eat different amounts or types of food, have different levels of exercise etc.

Page 34

1 Red
2 No genetic variation – they are a clone.
3 Reduction in allele variety so less variation for further breeding.
4 Crops designed for environmental conditions
  OR you get the features you want.
5 Human insulin is safe to use; it won't cause an allergic/immune response.

Page 36

1 At the bottom in the oldest rocks.
2 The genes/DNA of the skin cells.
3 Penicillin antibiotic won't kill these bacteria.
4 Simple life forms; insects; fish; reptiles; dinosaurs; birds; mammals.
5 Climatic change resulting in less food for them.

Page 38

1 Sperm
2 Son
3 a) black, b) black, c) brown
4 Yes, each child has a 50% chance of having cystic fibrosis.
5 One dominant and one recessive.

Page 39

1 Brain
2 Stimulates an egg to develop in the ovary.
3 Hormone 2 released mid-cycle from the brain.
4 Hormone O from the ovary.
5 Having several babies at once.

Topic test

1 a) (i) 23, (ii) 46
  b) mitotic division/mitosis
  c) (i) identical, (ii) clone (1 mark each)
2 a)–(iii), b)–(v), c)–(i), d)–(ii), e)–(iv) (1 mark each)
3 a) Dominant (1 mark)
  b)

	Mutant rat	Normal rat
	Rr	rr
Gametes	Ⓡ or ⓡ	ⓡ

	r	r
R	Rr	Rr
r	rr	rr

  50% of offspring are Rr; 50% are rr (3 marks)
  c)

	Mutant rat	Mutant rat
	Rr	Rr
Gametes	Ⓡ or ⓡ	Ⓡ or ⓡ

	R	r
R	RR	Rr
r	Rr	rr

  75% of offspring (6 rats) would not be killed by the poison. (3 marks)
  d) The rats which had the mutation would survive eating poison, (1 mark) reproduce and pass on their advantageous allele to their offspring. (1 mark)
4 a) Fertility drug stimulates ovulation. (1 mark)
  Oral contraception hormones (1 mark) prevent egg release from the ovaries. (1 mark)
  b) Multiple pregnancy. (1 mark)
5 a) (i) (human) insulin gene, (ii) bacterial, (iii) enzyme (3 marks)
  b) It will not cause an allergic reaction. (1 mark)
  It will not be broken down by the immune system. (1 mark)
6 a) Both parents are Cc. (2 marks)
  b) 75% chance that the child would not suffer from cystic fibrosis. (1 mark)

Answers

Module 5 – Metals

Check your understanding

Page 44

1 Relative atomic mass
2 Argon
3 70–75%
4 Any three from:
lower density; lower melting points; softer; not so tough/strong
5 Any three from:
much more reactive; corrode more rapidly; react with water to form soluble metal hydroxides and hydrogen; form white salts not coloured ones

Page 45

1 a) zinc + lead oxide → lead + zinc oxide
b) tin + aluminium chloride → tin chloride + aluminium
c) does not react
2 An ore is a rock from which useful substances can be extracted economically.
3 Carbon is more abundant and therefore much cheaper.

Page 46

1 It goes down (reduced).
2 It increases.
3 a) It would weaken the iron.
b) By reacting it with limestone to make slag.

Page 48

1 Because the ions are fixed in position and are not able to move.
2 Because the ions are separated and are free to move.
3 Bauxite (Al_2O_3)
4 a) The negative electrode (cathode)
b) Oxygen gas
c) Because they react with the oxygen and burn away to form carbon dioxide.

Page 49

1 Rusting
2 a) Chromium and nickel
b) Stainless steel
3 a) Magnesium and zinc
b) They must be higher in the reactivity series than iron.
c) Sacrificial protection
4 Because on exposure to air it gets covered with a layer of aluminium oxide. This layer seals the surface and prevents air and water from getting to the metal.

Page 51

1 7
2 Indicators
3 Water
4 a) potassium chloride
b) lithium nitrate
c) ammonium sulphate
5 An alkali is a base that is soluble in water.

Topic test

1 (i) chlorine (ii) mercury
(iii) iron (iv) carbon (1 mark each)
2 Mn and Ru (1 mark each)
3 Magnesium and carbon (1 mark each)
4 a) coke b) hot air
c) limestone d) iron ore (1 mark each)
5 a) (iii) b) (ii) c) (iv) d) (ii)
(1 mark each)
6 a) electrolysis (1 mark)
b) heating with carbon (1 mark)
c) electrolysis (1 mark)
7 Adding chromium (and nickel) to make stainless steel; using sacrificial protection with magnesium or zinc.
(Accept painting, cover in plastic, coat with oil.) (1 mark each)
8 Aluminium reacts with air/water (1 mark) to form a layer of aluminium oxide. (1 mark)
This layer seals the surface and prevents any air or water getting to it. (1 mark)
(This is an extended writing exercise.)
9 Any acid (e.g. sulphuric, citric etc.).
Accept just the word 'acid'. (1 mark)
10 a) (i) nitric acid
(ii) sodium hydroxide (or oxide)
(iii) potassium chloride (1 mark each)
b) Neutralisation (1 mark)
c) Water (1 mark)

Answers

Module 6 – Earth materials

Check your understanding

Page 54

1 a) limestone (calcium carbonate) → quicklime (calcium oxide) + carbon dioxide
b) quicklime (calcium oxide) + water → slaked lime (calcium hydroxide)
c) limestone (calcium carbonate) + sulphuric acid → calcium sulphate + carbon dioxide + water
2 Because it is too alkaline and could 'over compensate' making the water alkaline – which would be just as damaging.
3 Making iron in the blast furnace.

Page 55

1 A compound that contains carbon and hydrogen only.
2 Fractional distillation
3 Refinery gases; gasoline (petrol); kerosene (paraffin); gas oil (diesel); fuel oil; bitumen
4 Viscosity; boiling point; flammability; number of carbon atoms in the molecules; how volatile it is.
5 Those with a large number of carbon atoms (e.g. bitumen).
6 Cracking
7 Making polymers and as fuels.
8 Because plastics are not biodegradable.
9 Plastics can be burned (incinerated) producing useful heat but harmful gases. Some plastics can be recycled which reduces the use of natural resources.
10 Carbon dioxide and water (vapour)
11 Sulphur dioxide gas is produced. This causes acid rain which can kill wildlife.

Page 57 top

1 By volcanic activity.
2 During the first billion years of the Earth's existence.
3 Carbon dioxide (converted into oxygen by plants and absorbed into oceans where it became 'locked up' in carbonate). Water vapour (formed oceans as the planet cooled). Methane (reacted with oxygen to form carbon dioxide). Ammonia (reacted with oxygen and produced nitrogen).
4 Nitrogen (80%), oxygen (20%) and other gases (including carbon dioxide, water vapour and the noble gases).

Page 57 bottom

1 The inner core of the Earth has the highest density.
2 a) The main parts are (from the outside) the crust, the mantle, the outer core and the inner core.
b) The mantle has the largest volume.
3 The rocks are solid but the mantle flows like a very viscous liquid.

Page 59

1 They can be tilted, folded, faulted (fractured) or turned upside down.
2 Tectonic activity.
3 By the action of heat and pressure on existing rocks.

Page 60

1 The shape of the land masses and the fossil/rock record on the two continents.
2 a) Continental drift.
b) Because there was no way to explain why the continents could drift apart until about 50 years later.
3 a) The lithosphere is the crust and the upper part of the mantle.
b) Tectonic plates.
c) Convection currents in the Earth's core caused by radioactive processes.

Topic test

1 a) quicklime (1 mark)
b) cement (1 mark)
c) glass (1 mark)
d) building (1 mark)
2 a) and e) (1 mark each)
3 D and E (1 mark each)
4 a) (iv)
b) (iv)
c) (i)
d) (iv) (1 mark each)

Answers

5 a) Carbon dioxide and water vapour (1 mark each)
b) Oxygen and nitrogen (1 mark each)

6 a) Iron and nickel (1 mark each)
b) The crust (1 mark)
c) The crust, the mantle, the inner core (1 mark each)

7

Type of rock	How the rock is formed
Igneous	Heat and pressure on existing rocks
Metamorphic	Magma cooling
Sedimentary	Rock particles being deposited on the sea bed

(2 marks for 3 correct links, 1 mark for 1 or 2 correct links)

8 a) Faults (1 mark)
b) The movement of the Earth's crust. (1 mark)
c) The crust and the upper part of the mantle. (1 mark)

9 The plates move because of convection currents deep inside the Earth's core. (1 mark)
The heat that produces these currents (1 mark)
comes from radioactive processes inside the core. (1 mark)

Module 7 – Patterns of chemical change

Check your understanding

Page 63

1 Either harmful or irritant.
2 The minimum amount of energy needed to start a reaction.

Page 65

1 a) Corrosive, b) highly flammable.
2 A reactant: either sodium or water.
A product: either sodium hydroxide or hydrogen.
3

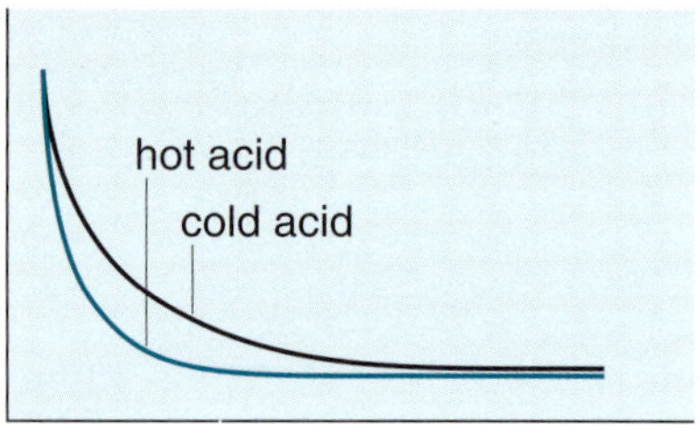

Page 66

1 Crush the marble into smaller pieces; heat the acid; use a concentrated solution of the acid.
2 The sticks have a much greater surface area than the log from which they were made so the air can get to them more easily and let them burn more quickly.
3 The low temperature in the refrigerator slows down the reactions that make the food turn bad.

Page 67

1 Catalysts produced by living organisms.
2 High temperatures destroy them.

Page 69

1 The energy supplied breaks the bonds in the reactants.
2 An exothermic process – energy is released when bonds are made.
3 An endothermic change because energy is being supplied.

Page 71

1 A reaction that can go in either direction depending on the conditions.
2 Add white anhydrous copper sulphate powder to the liquid. If the liquid contains water, the white powder will turn blue.
3 Pressure = 200 atmospheres; temperature = 450°C; catalyst of iron.

Answers

Page 72

1 Sulphur atoms are twice as heavy as oxygen atoms.
2 No. The correct answer should be 32 + 16 + 16 = 64.
3 7 (2 hydrogens + 1 sulphur + 4 oxygens).

Topic test

1 A substance that provides oxygen to allow other materials to burn more easily. (1 mark)
2 Corrosive (1 mark)
3 Reactants; collide; broken; activation; new; products. (1 mark each)
4 In concentrated acid the reacting acid particles are closer together and are more likely to collide with the magnesium (1 mark)
than in the dilute acid where they are further apart. (1 mark)
5 A catalyst is a substance that speeds up a chemical reaction (1 mark)
but remains chemically unchanged at the end. (1 mark)
6 The reactants are slowly being used up (1 mark)
so their concentration decreases as the reaction proceeds. (1 mark)
There are fewer reacting molecules left to react. (1 mark)
7 When one of the reactants has been completely used up. (1 mark)
8 High pressures increase the concentration of the gases (1 mark)
making the reaction go faster. (1 mark)
9 An enzyme is a catalyst produced by a living organism. (1 mark)
10 Any three from:
making beer and wine; making bread dough rise; making yoghurt; in detergents; in baby foods; making sweeteners etc. (1 mark each)
11 Low temperatures and therefore lower costs. (1 mark)
12 Exothermic reactions give out heat energy; (1 mark)
endothermic reactions absorb heat energy. (1 mark)
13 Exothermic: burning a fuel.
Endothermic: photosynthesis. (1 mark each)
14 Activation energy (1 mark)
15 Bubble the gas through lime water. (1 mark)
If it is carbon dioxide the lime water will turn cloudy. (1 mark)
16 A reaction that can take place in both the forward and the reverse direction depending on the conditions used. (1 mark)
17 $\rightleftharpoons$ (1 mark)
18 Advantage: better yields. (1 mark)
Disadvantage: excess may pollute water supplies. (1 mark)
19 Nitrogen from the air. Hydrogen from natural gas. (1 mark each)
20 450°C and 200 atmospheres (1 mark each)
21 Iron (1 mark)
22 High temperatures make the ammonia decompose back to nitrogen and hydrogen. (1 mark)
23 Ammonium nitrate or ammonium sulphate (1 mark)
24 NH_4NO_3: $M_r = 80$; % of N = 35%.
$(NH_4)_2SO_4$: $M_r = 132$; % of N = 21.21% (4 marks)

Module 8 – Structures and bonding

Check your understanding

Page 77

1 a)

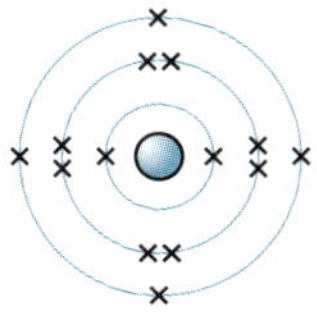

b)

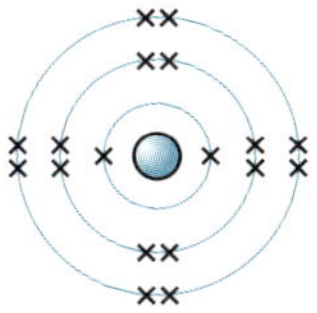

2 17 protons, 17 electrons, 18 neutrons

Page 81

1 (i) Electrons transfer from metal to non-metal elements.
(ii) Oppositely charged ions are formed.
(iii) Electrostatic forces of attraction hold the ions together.

Answers

2

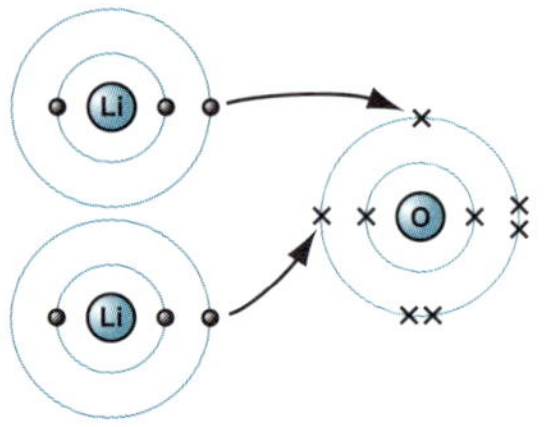

makes two Li^+ ions and one O^{2-} ion

Page 83

1 a)

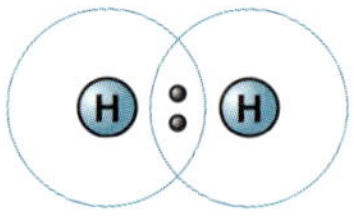

b)

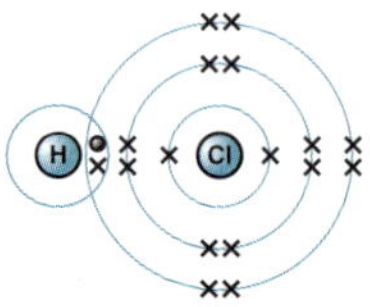

2 a) covalent, b) ionic, c) covalent
3 a) gas, b) solid, c) gas
4 They do not contain any free charged particles to carry the current.

Page 85

1 Groups
2 Reactivity increases down Group 1.
3 They are noble gases – all their energy levels are full.

Page 86

1 Electrolysis
2 Use an indicator paper, e.g. red litmus turns blue, or pH paper turns blue/violet.

Topic test

1 Nucleus (1 mark)
2 Energy levels (or shells) (1 mark)
3 Protons, p, 1+; neutrons, n, no charge (1 mark each)
4 Atoms of the same element with the same number of protons (1 mark)
but different numbers of neutrons. (1 mark)
5 It loses 1 electron. (1 mark)
6 It gains 2 electrons. (1 mark)
7

Particle	Number of			Atomic number	Mass number
	protons	neutrons	electrons		
N atom	7	7	7	7	14
B atom	5	6	5	5	11
Al atom	13	14	13	13	27
Ca^{2+} ion	20	20	18	20	40
F^- ion	9	10	10	9	19

(1 mark each)

8 a) Diagram showing one magnesium atom (2,8,2) (1 mark)
transferring its two outer electrons to two separate fluorine atoms (2,7). (2 marks)
b) ionic (1 mark)
c) solid (1 mark)
9 Group 5 because it has five outer electrons. (1 mark)
Period 3 because its electrons are in three energy levels (shells). (1 mark)
10 They all have free mobile electrons. (1 mark)
11 a) Reactivity increases down Group 1. (1 mark)
b) Reactivity decreases down Group 7. (1 mark)
12 All their energy levels (shells) are full. (1 mark)
13 Hydrogen, chlorine, sodium hydroxide (1 mark each)
14 It would slowly turn grey or darken. (1 mark)
15 a) Hydrogen gas: apply a flame; gas burns with a squeaky explosion. (1 mark)
b) Chlorine gas: add damp litmus paper; the litmus paper is bleached. (1 mark)
16 Any one from:
killing bacteria; making bleach; making hydrochloric acid; making PVC (1 mark)
17 a) NaBr (1 mark)
b) $MgBr_2$ (1 mark)
c) Na_2O (1 mark)
18 a) 1 × H, 1 × N, 3 × O (1 mark)
b) 2 × N, 8 × H, 1 × S, 4 × O (1 mark)
19 a) $CH_{4(g)} + \underline{2}\ O_{2(g)} \longrightarrow CO_{2(g)} + \underline{2}\ H_2O_{(l)}$ (1 mark each)
b) (g) = gas (1 mark)
(l) = liquid (1 mark)

Answers

Module 9 – Energy

Check your understanding

Page 92

1 Radiation
2 a) Black radiates away energy/black is a good emitter of radiation.
   b) White is a poor absorber of radiation/good reflector of radiation.
3 Double glazing, thicker window panes, thick curtains

Page 94

1 a) Loudspeaker (radio/electric keyboard/CD player etc.)
   b) Motor
   c) Lift
2 (Stored) energy due to raised position above ground.
3 joules (J), kilowatt-hours (kWh)
4 150 000 J/150 kJ

Page 95

1 Lamp B
2 0.02/2%
3 93.5 J

Page 97

1 a) Carbon dioxide/CO_2
   b) Sulphur dioxide/SO_2
2 Long-term radiation hazards/radioactivity lasts for thousands of years.
3 Electricity produced is too expensive.

Page 99

1 a) Hydroelectric
   b) Tidal
   c) Gas
2 Gas, hydroelectric
3 Satellite/any remote place/boat

Topic test

1 [A]–[G] [B]–[F] [C]–[H] [D]–[E] (1 mark each)
2 (i) steam, (ii) boiler, (iii) turbine, (iv) coal, (v) generator (1 mark each)
3 a) (i) transferred, (ii) cold/cooler/cool/colder, (iii) radiation/conduction, (iv) conduction/radiation, (v) slower/less, (vi) falls/drops, (vii) absorbs, (viii) faster, (ix) hotter/hot (1 mark each)
   b) £300 ÷ £60 (1 mark)
   5 years (1 mark)
4 a) Less wasted energy/more useful energy. (1 mark)
   b) Less fossil fuel/energy resources used to provide heating; less pollution caused in heating houses separately. (2 marks)
   c) Loss of heat to surroundings/water less hot as it reaches houses. (1 mark)
   d) Insulating pipes carrying hot water/using pipes of poor conducting material etc. (1 mark)
5 a) Gravitational potential energy (1 mark)
   b) 200 kJ/200 000 J (1 mark)
   c) 300 kJ/300 000 J (1 mark)
   d) 0.4/40% (1 mark)
6 a) 1525 (units) (1 mark)
   b) 1525 × 7p (1 mark)
   = £106.75 (1 mark)

Answers

Module 10 – Electricity

Check your understanding

Page 104

1 A diode only lets current flow in one direction.
2 a) 12 volts
   b) The resistance goes down as the temperature goes up.
3
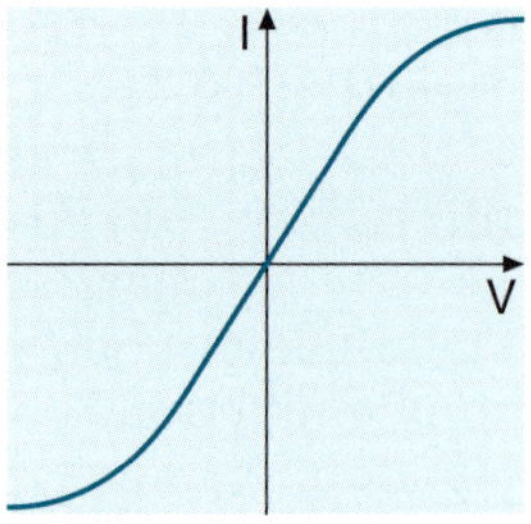

4 As light intensity increases, the current increases.

Page 105

1 The current creates a magnetic field around the wire. This magnetic field and that of the permanent magnets interact to produce a force on each side of the coil. The forces are in opposite directions so the coil turns.
2 It switches off the current faster than a fuse (it can also be reset).

Page 107

1 It is joined to the ground by a conductor (often a copper wire).
2 Repel – they have been given the same type of charge.
3 The balloon attracts and picks up the pieces of paper.

Page 110

1 Blue
2 It is a good conductor.
3 Plastic or rubber (they are good insulators).

4 a)
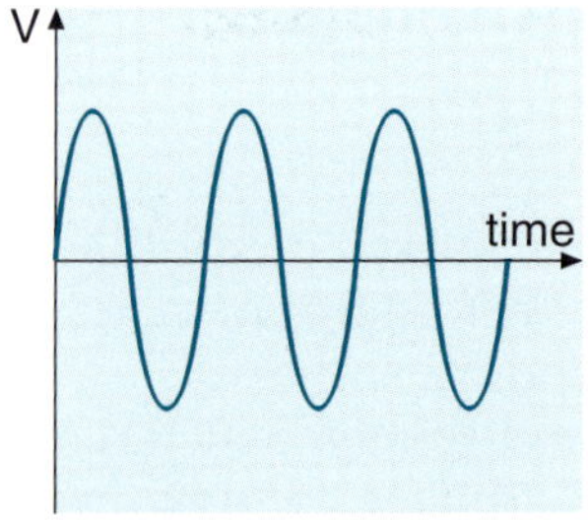

   b) If the frequency doubles the trace becomes
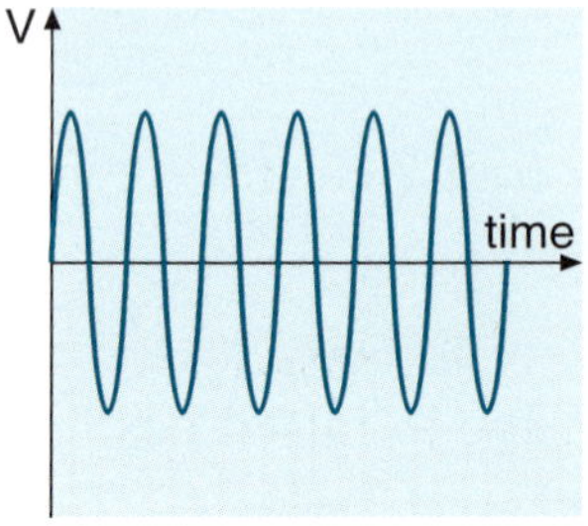

5 A current greater than the fuse rating causes the fuse to get so hot that it melts.

Page 112

1 3.6 watts
2 Use a stronger magnet; rotate the coil faster; add more turns to the coil; increase the area of the coil.
3 Substances (solid or gas) are deposited.
4 Electrical energy is transferred as heat.

Topic test

1 A – cell, B – ammeter, C- diode, D – resistor, E – LDR (1 mark each)
2 live; fuse; neutral; plastic; insulator; copper (1 mark each)
3 230 volts; 50 hertz; 1380 watts; 6 amps (1 mark each)
4 If the live wire touches the metal sole plate of the iron (1 mark)
  a large current flows from the live wire to earth (1 mark)
  causing the fuse to melt and break the circuit. (1 mark)

Answers

5 a) power = potential difference × current (1 mark)
= 230 × 8 (1 mark)
= 1840 watts (1 mark)

b) potential difference = current × resistance (1 mark)
230 = 8 × R (1 mark)
resistance = 29 ohms (approximately) (1 mark)

6 As the particles pass the metal grid they become charged the same as the grid. (1 mark)
The particles are repelled by the grid and attracted to the collecting plates. (1 mark)
The particles which stick to the plates are knocked off and removed. (1 mark)

7 a) Ammeter deflects two scale divisions to the right. (1 mark)
b) Ammeter deflects more than two divisions to the left. (1 mark)
c) Ammeter does not deflect at all. (1 mark)

Module 11 – Forces

Check your understanding

Page 115

1 Velocity includes direction.

2

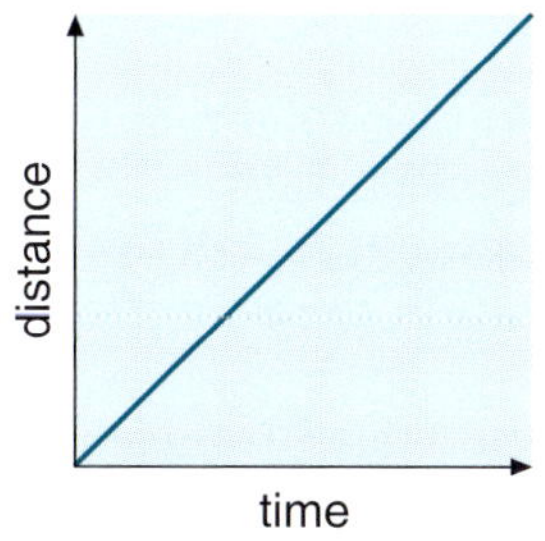

3 Constant velocity, then accelerates (constantly), then constant but higher velocity.

Page 118

1 8000 newtons
2 The force of gravity and air resistance are in the opposite direction to the ball's motion.
3 As the skydiver falls, he accelerates. The faster he goes, the greater the air resistance force. Eventually the air resistance upwards will equal the force due to gravity (weight) downwards.
4 The lorry has a greater mass than a car. So a greater force is needed to produce the same deceleration (F = m × a).

Page 120

1 The motorcyclist's reaction time (affected by alcohol, certain drugs, being tired); weather conditions (fog, ice, heavy rain etc.); condition of the motorcycle (worn tyres, worn brakes)
2 1000 joules
3 Elastic potential energy

Page 122

1 Gravity
2 So that they remain above the same point on the Earth's surface.
3 They are then able to scan the whole of the Earth's surface (at regular intervals).
4 As the comet approaches the Sun, the force increases; as the comet moves away from the Sun, the force decreases.

Page 125

1 A vast number of galaxies, at least a billion.
2 It will contract to become a white dwarf.
3 A neutron star.
4 Possibility that there was once water on the planet; fossil-like structures in rock.

Answers

Topic test

1 a) 750 metres (1 mark)
b) 450 seconds (1 mark)
c) D (1 mark)
d) Part B has a zero gradient (1 mark)

2 work = force × distance moved (1 mark)
= 2500 × 2.1 (1 mark)
= 5250 joules (1 mark)

3 Suitable examples include: communications; monitoring weather patterns; spying on countries (1 mark each)

4 galaxy; Universe; stars; gas; gravity (1 mark each)

5 Living organisms produce (chemical) changes to the atmosphere (1 mark) so it would be different from how it would have been with physical processes only. (1 mark)

6 Using radio telescopes (1 mark) to search for recognisable radio signals from space. (1 mark)

7 Steady (or constant) velocity to start (1 mark)
accelerates to a higher velocity (1 mark)
constant velocity (for a short time) (1 mark)
rapid deceleration (or slows down rapidly). (1 mark)

8 The forces are unbalanced. (1 mark)
The forces opposing motion (air resistance/drag) are bigger than the forward force. (1 mark)

9 The stopping distance increases. (1 mark)
Less friction between the tyres and the road. (1 mark)

10 weight = mass × gravitational field strength (1 mark)
= 70 × 1.6 (1 mark)
= 112 newtons (1 mark)

Module 12 – Waves and radiation

Check your understanding

Page 129

1 A transverse wave has the disturbance at right angles to the direction of travel.
Example from: water waves, light waves, radio waves, waves on a rope.
A longitudinal wave has the disturbance parallel to the direction of travel.
Example from: sound waves, waves along a stretched slinky spring.

2 Sound waves

3 a) 12 cm, b) 6 Hz, c) 72 cm/s

Page 130

1 Faster

2 It increases

3 Travelling along a normal

Page 131

1 Less diffraction occurs at shorter wavelengths, so waves do not diffract round the hill.

2 Increases

Page 134

1 Radio waves, microwaves, infrared, light, ultraviolet, X-rays, gamma rays

2 a) When ultraviolet radiation falls on the material, the radiation is re-emitted as visible light.
b) Used in washing powder brighteners, fluorescent lamps, security coding.

3 Labels 1 and 2 = electrical signal, copper cable; labels 3 and 4 = light signal, optical fibre; labels 5 and 6 = microwaves to satellite

Page 135

1 a) 6 cm, b) 80 Hz

2 Ultrasonic waves do not harm living cells but X-rays cause harm or kill cells.

Page 136

1 a) X and Y, b) X and Z, c) Y

Page 138

1 Cosmic rays, rocks, air, food, buildings

2 a) gamma, b) alpha

3 3 days

Answers

Topic test

1 a) 3 Hz (1 mark)
b) wave speed = frequency × wavelength (1 mark)
= 3 × 2.5 (1 mark)
= 7.5 (1 mark)
cm/s (1 mark)

2 Longitudinal: sound, spring (1 mark)
Transverse: water, light, radio, spring (1 mark)
Vacuum: light, radio (1 mark)

3 a) Bends away from the normal (1 mark)
b) Increases (1 mark)
c) Increases (1 mark)

4 a) Infrared, microwaves (1 mark)
b) X-rays (1 mark)
c) Infrared (1 mark)
d) Infrared (1 mark)
e) Ultraviolet (1 mark)

5 a) Sound with a frequency above the human hearing range. (1 mark)
b) Any two from:
pre-natal scans, cleaning delicate mechanisms or jewellery, detecting flaws in structures, echo sounding (1 mark each)

6 a) The radiation continually in the environment. (1 mark)
b) So the radiation emitted by the source can be worked out. (1 mark)
c) Emits alpha particles – rate goes down with thin paper. (1 mark)
Emits gamma rays – some radiation detected through lead. (1 mark)
Does not emit beta particles – count does not go down with 5 mm metal. (1 mark)

GLOSSARY/INDEX

Acceleration 114 – The rate at which velocity changes.

Acid 50 – A substance that dissolves in water to give a solution with a pH of less than 7.

Acid rain 28 – Atmospheric pollution caused by burning fossil fuels.

Activation energy 62 – The least amount of energy needed to start a chemical reaction.

Adaptation 22 – A feature of an organism that helps it to survive in a particular environment.

Aerobic respiration 6 – Respiration that takes place in the presence of oxygen.

Air resistance (drag) 116 – A force of friction.

Alkali 50 – A base (metal oxide or hydroxide) that dissolves in water to form a solution with a pH greater than 7.

Allele 32 – One of the different forms of a particular gene, e.g. the gene for coat colour can be B allele coding for black or b allele coding for brown.

Alloy 49 – A mixture of metals (or of carbon with a metal). Alloys can have properties different from the parent metal(s).

Alternating current (a.c.) 109 – An electric current that reverses direction regularly.

Alveoli (alveolus) 5 – Microscopic air sacs in the lungs which act as the surface for gaseous exchange.

Amino acids 3 – The breakdown products of the digestion of proteins, and the building blocks for making new proteins.

Amplitude 128 – The maximum disturbance of a wave from the equilibrium prosition.

Anaerobic respiration 6 – Respiration that takes place in the absence of oxygen.

Analogue signals 133 – Signals which vary continuously in amplitude and/or frequency.

Anode 47 – The positively charged electrode.

Antibodies 10 – Proteins made by some white blood cells which destroy some micro-organisms.

Aorta 7 – The main artery of the body which carries oxygenated blood away from the heart.

Arid 22 – An arid area is very dry and has a shortage of water.

Artery 7 – A blood vessel that carries blood away from the heart.

Asexual reproduction 31 – Reproduction that does not involve the formation and fusion of gametes. The offspring and parent have identical genetic information and no variation.

Atmosphere 56 – The layer of gases around the Earth.

Atom 75 – The smallest part of an element that can exist. Atoms have a nucleus consisting of protons and neutrons around which are shells of electrons.

Atomic number 75 – The number of protons present in an atomic nucleus (and the number of electrons present in the neutral atom).

Atria (atrium) 7 – The upper chambers of the heart which receive blood from the veins.

Background radiation 137 – The low-level radiation that is always present in the environment.

Bacteria (bacterium) 9 – Single-celled organisms made up of cytoplasm and a membrane surrounded by a cell wall, with genetic material not in a distinct nucleus.

Base 50 – An oxide or hydroxide of a metal.

Battery 101 – Two or more electrical cells joined together.

Bauxite 47 – The main ore of aluminium containing aluminium oxide.

Bile 3 – A liquid produced by the gall bladder that breaks up fat into droplets.

Biodegradable 55 – Materials which can be broken down (decomposed) by bacteria.

Biomass 24 – The mass of dry matter in an organism or at one feeding level of the food web.

Blast furnace 46 – The industrial method used for extracting metals such as iron from their ores.

Bonds 78, 81 – The forces that hold atoms together.

Braking distance 118 – How far a car travels before stopping, after the brakes have been applied.

Camouflage 22 – Colouration or pattern which enables an animal to remain hidden in its environment.

Capillaries 8 – Tiny, thin-walled blood vessels which join arteries and veins and allow substances to be exchanged through their walls.

Carbon cycle 26 – The recycling of carbon in the environment.

Carnivore 24 – An animal that eats other animals.

Catalyst 64 – A substance that changes the speed of a reaction, but remains unchanged after the reaction.

Cathode 47 – The negatively charged electrode.

Cell (biological) 1, 12 – The smallest part of an animal or plant.

Cell membrane 1, 12 – The very thin membrane on the outside of the cell cytoplasm that controls the movement of substances into and out of the cell.

Cell wall 12 – The outer part of a plant cell that gives strength and shape to the plant cell.

Cellulose 14 – The material that makes up the cell wall of plant cells.

Charge 75, 106 – A feature of atomic particles. Protons and electrons have a charge. Electrons have a negative charge and protons have a positive charge. Opposite charges attract; like charges repel.

Chlorophyll 13 – A green pigment found in the leaves and stems of plants. It traps light energy for use in photosynthesis.

Chloroplast 12 – That part of the plant cell containing chlorophyll.

Chromosome 32 – A single long molecule of DNA which looks like a thread in the nucleus. The DNA carries the genetic code information as a line of genes.

Circuit breaker 105 – A device used to break a circuit very quickly.

Clone 33 – An organism produced from one parent. The clone is genetically identical to the parent.

Community 23 – All the plants and animals living in one habitat.

Competition 23 – Occurs when plants or animals are fighting for food, light or space.

Compound 75 – A substance which contains two or more elements chemically joined together.

Continental drift 59 – The movement of the Earth's tectonic plates which has resulted in the continental masses moving apart.

Convection 91 – The transfer of heat energy in a liquid or gas (fluid) caused by differences in density. Warmer, less dense fluids rise as cooler, more dense fluids sink.

Core (Earth) 57 – The innermost part of the Earth.

Count rate 138 – The number of radioactive emissions detected in a certain time.

Covalent bond 81 – The bonding of atoms caused by the sharing of pairs of electrons in their outer electron shells.

Cracking 55 – A form of thermal decomposition in which large hydrocarbon molecules are broken down into smaller ones.

Crude oil 54 – A mixture of substances, most of which are hydrocarbons, formed by the anaerobic decomposition of marine organisms over a long period of time.

Crust 57 – The surface layer of the Earth.

Deforestation 27 – The removal of a large number of trees from an area by cutting down or burning.

Deposition 58 – The laying down in water of a layer of rock fragments.

Diffraction 131 – The spreading out of waves as they go through a gap or pass by an object.

Diffusion 2 – The movement of particles from regions where they are in high concentration to regions where they are in lower concentration.

Digital signals 133 – Signals coded as a set of pulses either on or off.

Diode 104 – An electrical component that only conducts electricity in one direction.

Direct current (d.c.) 109 – An electric current that always flows in the same direction.

Dominant allele 32, 37 – The allele which controls an observable characteristic in the offspring when it is present on only one chromosome.

Drug 19 – A chemical that changes the way our body works or the way we react to things.

Earthing 107, 109 – A metal object is earthed when a low resistance conductor joins the object to the Earth's surface.

Efficiency 95 – The ratio of useful energy transferred by a device, to total energy supplied to the device.

Elastic potential energy 120 – The energy stored in an elastic object when work is done to change its shape.

Electric current 110 – A flow of electrons or ions.

Electrode 47 – A negatively or positively charged conductor.

Electrolysis 47 – The process of splitting up a chemical compound using an electric current.

Electromagnet 104 – A device made by passing an electric current through a coil of wire.

Electromagnetic waves 132 – A range of transverse waves that travel through a vacuum (space) with a common speed, but with different frequencies and wavelengths.

Electrons 75 – Negatively charged particles orbiting in shells around the atomic nucleus.

Element 75 – A substance made up of atoms that contain the same number of protons, i.e. a substance containing only one type of atom. An element cannot be broken down into anything simpler by chemical means.

Endothermic reaction 68 – A reaction in which heat energy is transferred from the surroundings, because more energy is needed to break the existing bonds in the reactants than is released when new bonds are made in the products.

Energy levels (shells) 75 – Regions outside the nucleus where electrons are found.

Environment 22 – The surroundings and conditions that affect the growth and behaviour of plants and animals.

Enzyme 3, 66 – Protein that acts as a catalyst to change large insoluble molecules to small soluble ones.

Erosion 58 – The wearing away of the Earth's surface and transporting of the dust and rocks.

Exothermic reaction 67 – A reaction in which heat energy is transferred to the surroundings because more energy is given out making the new chemical bonds in the products than is taken in to break the existing bonds in the reactants.

Extinct 35 – A description of an organism which is no longer living today, but which lived in the past.

Faults 58 – Fractures in the Earth's crust caused by movement of the crust.

Fermentation 66 – The changing of glucose into ethanol (alcohol) and carbon dioxide by the action of enzymes in yeast.

Fertilisation 31 – The joining of the nuclei of the male and female sex cells.

Fertiliser 27, 70 – A chemical used to increase the growth rate and quality of plants.

Food chain 24 – A diagram which shows the feeding relationships of some organisms in an ecosystem.

Food web 25 – A diagram which shows the many feeding relationships in an environment.

Fossils 35 – The remains or imprint of plants or animals that lived millions of years ago. The remains may have been mineralised and turned to stone.

Fossil fuels 56 – Non-renewable fuels – oil, coal and natural gas, which were formed millions of years ago and therefore contain 'locked-up' carbon.

Fractional distillation 54 – A method of separating liquids whose boiling points are close together. The process is used to separate the different substances in crude oil.

Free electrons 83 – The electrons in metals that move around inside the metal and do not remain in orbit around a nucleus. The presence of these free electrons allows the metal to conduct electricity and heat.

Frequency 128 – The number of waves passing a certain point in one second. Unit is hertz (Hz).

Friction 116 – A force that tries to stop or does stop movement.

Fuse 108 – A wire designed to melt if too large a current flows through it.

Galaxy 123 – A vast number of stars held together by gravity.

Gamma rays (radiation) 132, 137 – Electromagnetic waves with very high frequency and low wavelength. Emitted by the unstable nuclei of radioactive substances.

Gene 32 – A length of DNA on a chromosome which controls the development of a characteristic.

Generator 98 – A device that transfers kinetic energy as electrical energy.

Geostationary orbit 122 – The orbit of a satellite that takes 24 hours to complete.

Geothermal energy 97 – The heat produced by the Earth by radioactive decay within rocks.

Global warming 28 – The international problem caused by increased levels of carbon dioxide and methane in the atmosphere resulting in a small increase in the Earth's temperature. The consequences are climate changes and flooding.

Gravitational potential energy 93 – The energy stored by an object due to its vertical height above the Earth's surface.

Gravity 122 – The force of attraction that acts between all bodies.

Greenhouse effect 96 – The effect in the atmosphere of heat energy being absorbed by gases such as carbon dioxide and methane.

Group (chemical) 78 – A vertical column of elements in the Periodic Table having similar chemical properties due to the atoms of the elements having the same number of electrons in their outer shells.

Haber process 70 – The process used for the manufacture of ammonia.

Halide 86 – The compound formed when a halogen reacts with another element.

Half-life 138 – The time taken for the count rate of a radioactive substance to fall to half of its original value.

Halogens 84 – The name given to the elements in Group 7 of the Periodic Table.

Herbivore 24 – An organism that eats only plants.

Hormone (plant) 15 – A chemical (protein) in plants which can affect growth and acts under the effect of water, light and gravity.

Hormones (in fertility) 39 – Chemicals produced in the brain which control the menstrual cycle, or artificial forms.

Hydrocarbon 54 – A compound containing only hydrogen and carbon.

Hydroelectric power 98 – Electrical power generated by the flow of water from a higher to a lower level.

Hydrogen ion 50 – An ion (H^+) present in all acids.

Hydroxide ion 50 – An ion (OH^-) present in all alkalis.

Induced current 111 – The current produced when a wire that is part of a complete circuit cuts through a magnetic field.

Ion 47, 110 – An atom or group of atoms which have lost or gained electrons to become positively or negatively charged.

Ionic bond 78 – The electrostatic attraction between opposite charges responsible for holding metal and non-metal elements together in a compound. The ions are formed when the metal atoms transfer electrons to the non-metal atoms in order to achieve full outer electron shells.

Ionise 137 – To remove or add electrons to atoms or groups of atoms so giving them positive or negative charges.

Iris 17 – A ring of muscle which controls the amount of light entering the eye.

Isotope 75, 136 – Atoms of the same element which contain different numbers of neutrons.

Joule 93 – The unit of energy and work.

Kilowatt-hour 94 – A unit of electrical energy.

Lens (eye) 17 – A transparent structure within the eye that is flexible and helps light to form a sharp image on the retina during focusing.

Light dependent resistor (LDR) 103 – An electrical component whose resistance goes down when a bright light shines on it.

Limestone 53 – A sedimentary rock (mainly calcium carbonate) formed millions of years ago from the shells of small creatures.

Lithosphere 59 – The outer shell of the Earth made from the crust and the upper part of the mantle.

Longitudinal wave 128 – A wave in which energy is transferred in the same direction as the vibrations causing the wave.

Mantle 57 – The layer of the Earth between the crust and the core.

Mass number 75 – The total number of protons and neutrons in an atomic nucleus.

Metamorphic rocks 58 – Rocks formed from rocks which became buried deep underground and had their structure changed by high temperatures and/or high pressures.

Micro-organisms 9, 26 – Organisms such as bacteria and fungi which can only be seen with a microscope. Bacteria and fungi are together known as decomposers.

Milky Way galaxy 123 – The galaxy that contains our Solar System.

Mixture 54 – Two or more substances, which are usually easy to separate.

Molecule 81 – A particle containing atoms of the same or different elements bonded together. The smallest part of an element or compound that can take part in a chemical reaction.

Mutation 36 – A sudden change occurring in a gene or a chromosome which results in a change in a feature of the organism.

Natural selection 36 – The process by which a beneficial characteristic with a greater survival value is selected and passed on so that the proportion of the allele increases. Natural selection leads to evolution.

Neutralisation 50 – A reaction between an acid and a base or a carbonate.

Neutron 75 – A particle with no electrical charge found in the nucleus of most atoms. Its mass is similar to that of a proton.

Noble gases 84 – The name given to the elements in Group 0 of the Periodic Table.

Non-metals 78, 81 – Elements in the Periodic Table which usually have low melting points and boiling points, are poor conductors of electricity and heat, and as solids are brittle.

Nucleons 136 – The protons and neutrons in the nucleus of an atom.

Nucleon number 136 – The total number of protons and neutrons in an atomic nucleus.

Nucleus (atom) 75 – The central part of an atom that contains positively charged protons and uncharged neutrons.

Ohm 103 – The unit of electrical resistance.

Optical fibre 131 – A tube of plastic or glass that allows light to travel from one end to the other by repeated total internal reflection.

Orbit 121 – The path taken by an object which goes around another object.

Ore 45 – A mineral or mixture of minerals from which a metal can be extracted in economically viable amounts.

Osmosis 14 – Diffusion of water from a dilute to a more concentrated solution, through a selectively permeable membrane.

Oxidation 46 – The process in which an element or compound gains oxygen.

Period (chemical) 43 – A horizontal row of elements in the Periodic Table.

Periodic Table 43 – The arrangement of the elements in order of increasing atomic number.

Pesticide 27 – A chemical used to kill unwanted plants or animals regarded as pests.

pH 66 – A scale used to measure acidity and alkalinity.

Photosynthesis 13 – The production of glucose by green plants from carbon dioxide and water using sunlight energy and giving out oxygen.

Plastic 55 – The common name for some types of polymer.

Polymer 55 – A long chain molecule made up of many smaller molecules called monomers.

Population 23 – The numbers of one species of plant or animal living in one habitat.

Power 93 – The rate of transferring energy (doing work). The unit is the watt (W), where 1000 watts = 1 kilowatt.

Predator 23 – An organism that kills and eats another organism (known as the prey).

Prey 23 – An organism that is caught and eaten by a predator.

Products 62 – The new materials produced as a result of a chemical reaction.

Proton 75 – A positively charged particle found in the nucleus of an atom.

Pyramid of biomass 25 – A diagram which shows by the size of the block (quantitatively) the biomass at each feeding level.

Radiation (thermal) 91 – The transfer of heat energy by electromagnetic waves travelling through space.

Radioactivity 137 – The random emission of energy from an atomic nucleus as the result of the breakdown of unstable nuclei.

Rate of reaction 63 – The speed at which a chemical reaction progresses.

Reactants 62 – The starting materials in a chemical reaction.

Reactivity series 44 – A list of metals arranged in order of their chemical reactivity. The most reactive metals are at the top of the list.

Recessive allele 37 – This allele must be present on both chromosomes of the pair to show in the offspring.

Red giant 124 – A relatively cool giant star.

Reduction 45 – The process in which an element or compound loses oxygen.

Reflex action 17 – A rapid automatic response to a stimulus, during which nerve impulses are sent by receptors through the nervous system to effectors.

Refraction 129 – The change in direction of a wave as it passes from one medium into another due to the change in the speed of the wave.

Relative atomic mass 43 – The average mass of an atom of an element on a scale on which the mass of a hydrogen atom = 1 or the mass of the ^{12}C isotope of carbon = 12. It takes into account the relative abundance of different isotopes with different mass numbers.

Relative formula mass 72 – This is found by adding together the relative atomic masses of all the elements in the chemical formula of the substance.

Respiration 6 – A chemical reaction that occurs in all living cells to oxidise glucose to carbon dioxide and water and to release energy.

Retina 17 – A light-sensitive layer at the back of the eye which sends impulses to the brain by the optic nerve.

Reversible reaction 69 – A reaction that can proceed in either direction depending on the reaction conditions. Reactants can be changed into products which in turn can be changed back into reactants.

Sacrificial protection 49 – Used to reduce the rusting of iron by attaching a more reactive metal such as magnesium or zinc.

Salt 50 – A substance formed when an acid reacts with a base.

Satellite 122 – An object in orbit around the Earth.

Sclera 16 – The tough outer layer of the eye.

Sedimentary rocks 58 – Rocks formed by deposition, burial and compression of weathered rock fragments or the shells of dead animals. They can also be formed by the precipitation of calcium carbonate, usually in warm, shallow seas.

Seismic waves 135 – Waves created by the vibrations caused by earthquakes.

Selective breeding 34 – Choosing a desirable characteristic and breeding plants or animals for many generations until they show this feature.

Sexual reproduction 31 – Involves two parents who each produce sex cells that join together. There is genetic variation in the offspring.

Shells (energy levels) 75 – Regions outside the nucleus where electrons are found.

Slag 46 – The substance formed in the blast furnace from impurities in the iron.

Star 124 – A huge mass that produces its own light and heat.

Stomata 12 – Small holes in the surface of leaves, usually on the underside, which are controlled by guard cells.

Stopping distance 118 – Thinking distance + braking distance

Tectonic plates 59 – The separate slow-moving adjacent sections of the Earth's lithosphere that move because of convection currents within the Earth's mantle caused by the natural radioactive processes within the Earth.

Terminal velocity 118 – The constant speed reached by a moving object when the forces (in the direction it is moving) are balanced.

Thermal decomposition 53 – The breaking down of a compound by the action of heat.

Thermistor 103 – An electrical component whose resistance goes down when it gets hot.

Thinking distance 118 – How far a car travels during the driver's reaction time, before the brakes are applied.

Total internal reflection 130 – This happens when light moving through a transparent material into air hits the surface of the material at an angle bigger than the critical angle.

Toxic 27 – Poisonous; harmful to living organisms.

Transformer 112 – A device that increases or decreases an alternating voltage.

Transition elements (metals) 43 – The name given to the elements in the Periodic Table between Groups 2 and 3.

Transpiration 13 – Loss of water from the leaves of a plant by evaporation through the stomata.

Transverse wave 127 – A wave in which the energy is transferred at right angles to the vibrations causing the waves.

Ultrasound 134 – Sound waves with frequencies above the upper limit of human hearing (above 20 000 Hz).

Universe 123 – Made up of countless galaxies. It is everything there is.

Velocity 114 – Speed in a given direction.

Ventilation 5 – Movement of air into and out of the lungs during breathing.

Wavelength 128 – The distance from a particular point on one wave to the same point on the next wave.

Weathering 58 – The chemical, physical or biological action by which rocks are broken down into rock fragments.

Weight 117 – The force due to gravity on an object. Measured in newtons (N).

White dwarf 124 – A small, very dense, bright star.